Encyclopaedia of Life Science

Encyclopaedia of Life Science

VOL. 4

B.A. Mannan

ANMOL PUBLICATIONS PVT. LTD.
NEW DELHI - 110 002 (INDIA)

ANMOL PUBLICATIONS PVT. LTD.
H.O.: 4374/4B, Ansari Road, Darya Ganj,
New Delhi-110 002 (India)
Ph.: 23278000, 23261597
B.O.: No. 1015, Ist Main Road, BSK IIIrd Stage
IIIrd Phase, IIIrd Block,
Bangalore - 560 085 (India)
Visit us at: www.anmolpublications.com

Encyclopaedia of Life Science

First Published, 2007

ISBN 978-81-261-3408-3 (Set)

PRINTED IN INDIA

Printed at Amit Enterprises, Delhi.

Contents

Preface

'Life Science' is a branch of science, under which we study about the living beings in detail. We study about the body structure; all the organs of the body and their functions. The body consists of several organs, which are made of Tissues, while, Tissues are made of cells. Plants are very useful for living organisms as they supply oxygen for the respiration and they are the food-source for the primary consumers. Plant survival and growth is critically influenced by abiotic factors including water, air and light. The effect of the local environment on plant growth also accounts for much of variation in adult form within a species. Plants have green pigments, which are quite useful in the process of photosynthesis. In the presence of light, plants make their food with the help of minerals, water and carbondioxide, in their green leaves, in the form of carbohydrates. The food prepared in green leaves is transported to different parts of the plant by phloem.

Similarly, animal life is a significant feature on earth. All the functions of the human beings and animals, including birds and insects, are performed in their cells. Cell is the structural unit of any living body. This was first discovered by scientist, Robert Hook. The cell has several parts, such as nucleus, centrosomes, mitochondria, plastids, cell-walls. The cell wall gives the mechanical support to the cell and cell-walls are accountable for the transportation of the material in and out of the cells.

Biologists agree that all living things arose through long periods of change shaped by physical and chemical processes. Approximately, there are two million species in the living world. In changing environment, a few species are eliminated during un-favourable conditions on the basis of the principle of the 'survival of the fittest,' which is also known as natural selection.

Science can strive to grasp nature's faculty, but science is socially embedded and all scientists record prevailing certainties. Darwin himself, in the closing lines of 'The Origin of Species', mentioned about Victorial social preference. Life's pathway certainly includes many features, predicable from laws of nature. But laws of nature do not tell us why we have crabs and snails at all, why insects rule the multicellular world, and why vertebrates rather than persistent algal mats exist as the most complex form of life on earth.

Bacteria too represent the great success story of life's pathway. They occupy a wider domain of environments and span a border range of biochemistries than any other group. One might conclude that complexification of life as a whole represents a pseudo-trend based on constraints at the wall, but still holds that evaluation continues, within a particular group.

This eminent work, titled; *Encyclopaedia of Life Science* would be able to make new waves in the discipline. Expectedly, these books would prove to be the best source of study in the area of Life Science.

Editor

1

Role of DNA

The nucleic acids were discovered over 100 years ago, but their role in heredity and in the control of cellular metabolism has been described only during the past 50 years. There are two nucleic acids, deoxyribonucleic acid (DNA) and ribonucleic acid (RNA). The DNA is the predominant nucleic acid in the nucleus, while the RNA is the predominant nucleic acid in the cytoplasm, although all cytoplasmic RNA is of nuclear origin.

Significance of Deoxyribonucleic Acid

In 1968, *F. Miescher* extracted from the nuclei of pus cells (white blood cells) a substance which he called *nuclein*. However, Miescher was not aware that the substance he isolated was the hereditary material. It was later on renamed *deoxyribonucleic acid* (DNA) and is now recognised as being the chemical structure that stores the cell's hereditary information.

In eucaryotic cells, two distinct types of DNA are present. The nuclear DNA of eucaryotes differs fundamentally from other DNA in that it is complexed with certain proteins called *histones*, with *non-histone* proteins, and with *small amounts of RNA* to form the structures called *chromosomes*.

The DNA of self-replicating organelles such as mitochondria and chloroplasts resembles closely that of procaryotes (a) in its relatively smaller size, (b) in not being found in combination with histone proteins, and (c) in its circular form.

Molecular Structure of DNA

DNA is composed of:

1. *A Pentose Sugar, called deoxyribose ("de" = lacking, "deoxy"=* lacking in an oxygen atom found in the ribose at no. 2 position).
2. *Nitrogen Bases:* There are four heterocyclic bases in DNA. Two of the bases are *purines:* Adenine (A) and Guanine (G), and two are *pyrimidines:* Cytosine (C) and Thymine (T). As separate chemicals these bases exist largely in their *enol* forms; but when these purines and pyrimidines are bound into nucleic acids, they are in their *keto* forms. In the DNA molecule, two hydrogen bonds exist between adenine and thymine and three hydrogen bonds join guanine and cytosine.
3. *Phosphoric Acid:* or H_3PO_4 links the nucleotides by joining the pentoses of two consecutive nucleotides with an ester-phosphate bond. This bond links carbon 3' in one nucleoside with carbon 5' in the next. In this way phosphoric acid uses two of its three groups. The remaining acid group enables the molecule to form ionic bonds with basic proteins (histones and protamines). This group makes nucleotides highly basophilic, *i.e.,* they stain readily with basic dyes.

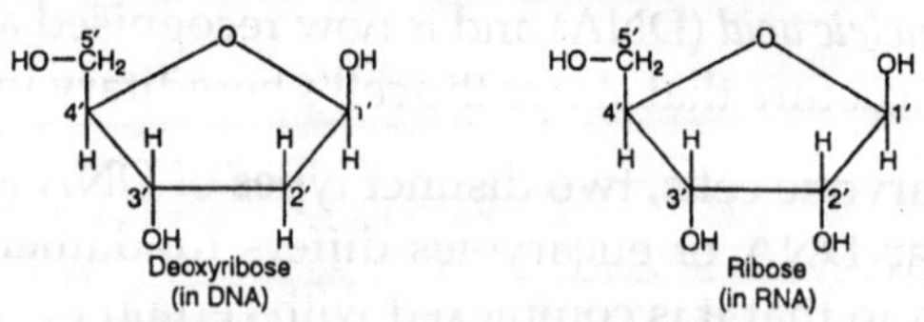

The Pentose Sugars of Nucleic Acids.

In DNA, each of the bases is bound to a deoxyribose sugar. Linkage between a purine base and the sugar occurs between group 9 (base) and group 1' (sugar), while pyrimidines link at group 3. The resulting combinations of base and sugar are known as *nucleosides* and are designated as deoxyadenosine, deoxyguanosine, deoxycytidine and thymidine respectively.

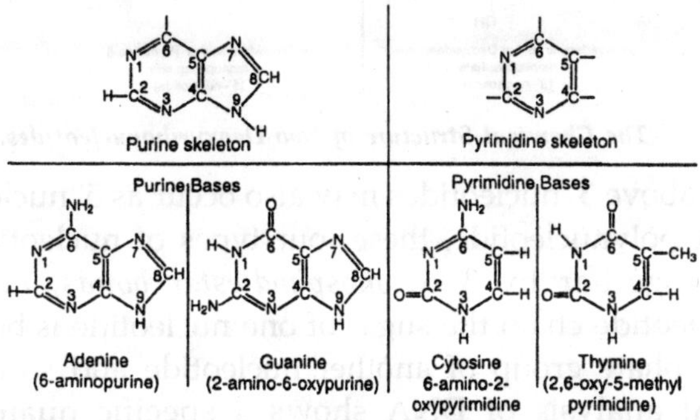

The Purine and Pyrimidine Basés of DNA.

A combination of base, sugar and a phosphate group is called *nucleotide.* The combination with the deoxyribose part of the nucleoside may occur at group 3' or group 5', and the resulting nucleotides are called 3' nucleotides and 5' nucleotides respectively. The four nucleotides of DNA, called *deoxyribonucleotides,* are as follows:

(i) Deoxyadenylic acid or Deoxyadenosine-3'-monophosphate.

(ii) Deoxyguanylic acid or Deoxyguanosine-3'-monophosphate.

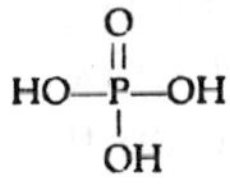

Phosphoric Acid (H_3PO_4)

(iii) Deoxycytidylic acid or Deoxycytidine-3'-monophosphate.

(iv) Deoxythymidylic acid or Dcoxythymidine-3'-monophosphate.

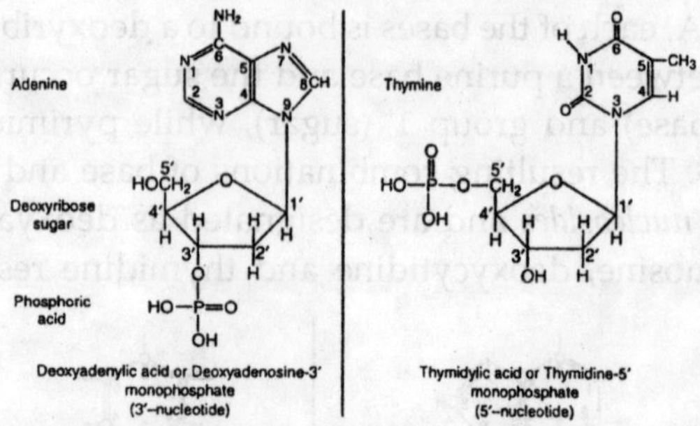

The Chemical Structure of two Deoxyribonucleotides.

The above 3' nucleotides may also occur as 5' nucleotides. In DNA polynucleotide, these four types of nucleotides are linked together by 3'-5' *phosphodiester bonds, i.e.,* in a polynucleotide chain the sugar of one nucleotide is bound to the phosphate group of another nucleotide and so on. The chemical analysis of DNA shows a specific quantitative relationship between the different nucleotides. For each nucleotide of adenine there is one of thymine (A = T), and for each nucleotide of guanine there is one of cytosine (G = C).

Watson and Crick Model of DNA

In 1950, *E. Chargaff* pointed out that in DNA, the molar amount of Adenine equals the molar amount of thymine and the molar amount of guanine equals the molar amount of cytosine. This observation has become a universal biological law which holds true for every form of DNA found in nature. While the work on the base composition of DNA was in progress in Chargaff's laboratory, R. *Franklin* and *M. Wilkins* used X-ray diffraction to obtain precise measurements of DNA. In 1951, *J.D.Watson* and *F.H. Crick* met in Cambridge (England) and decided to collaborate on the study of DNA.

In 1953, *J.D. Watson* and *F.H. Crick* proposed a model of the structure of the DNA molecule that has been widely accepted. For this work these two men shared with *Wilkins,* the 1962 Nobel Prize in medicine and physiology. According to this model, DNA usually consists of two very long, thin,

polymeric, anti-parallel chains twisted about each other in the form of a regular *double helix* . Each chain is a polynucleotide or a collection of many nucleotides in which the sugar of one nucleotide is linked by a phosphate group to the sugar of the adjacent nucleotide.

The diameter of the helix is about 20 Å and each chain makes a complete turn every 34 Å. As there are 10 nucleotides on each chain for every turn of the helix, the distance per nucleotide base is thus 3-4 Å.

There are four nucleotides, each of them containing a deoxyribose sugar (D), a phosphate group (P) and a purine or pyrimidine base. There are two *purines,* adenine (A) and guanine (G), and two *pyrimidines,* cytosine (C) and thymine (T). In the polynucleotide chain the joining together of the sugar and phosphate groups always involves the same chemical groups. Hence, this part of the molecule, called the backbone, is very regular.

In contrast, the order of the purine and pyrimidine bases along the chain is highly irregular and varies from one molecule to another. Both the purine and pyrimidine bases are flat, relatively water-insoluble molecules which tend to stack above each other perpendicular to the direction of the helical axis.

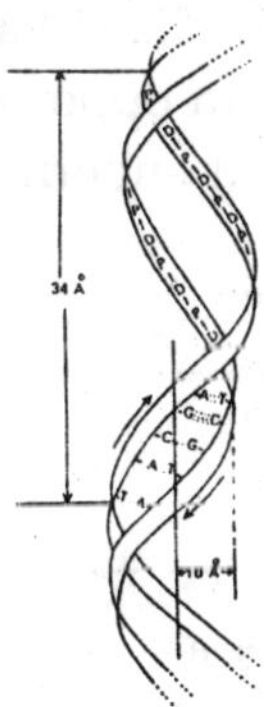

The DNA Double Helix, Watson and Crick Model.

The two polynucleotide chains are joined together by *hydrogen bonds* between the pairs of bases . According to the base-pairing rules *adenine* (purine) is always joined to *thymine* (pyrimidine) by two hydrogen bonds, and *guanine* (purine) is always bounded to *cytosine* (pyrimidine) by three hydrogen bonds. A hydrogen bond takes place from the sharing of a single hydrogen atom by two atoms, one of which acts as a hydrogen donor and the other as a hydrogen acceptor. These donor and acceptor atoms are found in each of the bases of DNA. Although the hydrogen bonds are weaker than the ordinary chemical bonds, the great number of these bonds contributes to the stability of the double helical structure.

Thus, in the DNA double helix hydrogen bonds exist only between adenine and thymine (A and T), and between guanine and cytosine (G and C) because other arrangements distort the helical structure of DNA. Hence, only these arrangements are possible for two purines (A and G) would occupy too much space to allow a regular helix and, correspondingly, two pyrimidines (C and T) would occupy too little. The existence of the base pairs means that the sequence of bases and, hence of nucleotides along the two chains are not identical but *complementary*. For example, if we have a sequence ATGCT on one chain, the opposite chain must have the sequence TACGA. According to *Watson,* a stereochemical consequence of the formation of the A...T and G...C base pairs is that the two polynucleotide chains run in opposite directions. Thus, if the helix is inverted by 180°, it superficially looks the same.

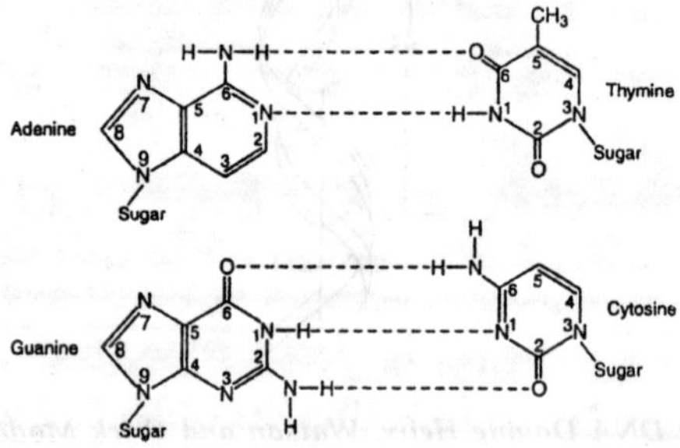

The Pairing of Purine and Pyrimidine bases by Means of Hydrogen Bonds.

An important chemical feature of the DNA structure is the position of the hydrogen atoms in the purine and pyrimidine bases, for the H atoms are found at precise or fixed locations. The N atoms attached to the purine and pyrimidine rings are usually in the *amino* (NH_2) form and only very rarely assume the imino (NH) configuration. Similarly, the *oxygen atoms* attached to the C-6 atoms of guanine and thymine normally have the *keto* (C = O) form, and only rarely take up the *enol* (COH) configuration. These relatively stable locations are quite essential to the biological function of DNA.

In contrast to the ratios of A/T and G/C, which must always be 1-1 to satisfy the base-pairing rules, there are wide variations in the A + T/ G + C contents of different DNA molecules. Higher plants and animals all have an excess of A + T over G + C in their DNA, whereas, among the viruses, bacteria and lower plants, there is much more variation; and both A + T-rich and G + C-rich species occur. However, these variations are not purely random, and the base ratios of taxonomically related organisms are quite similar.

DNA molecules with a high G + C content are more resistant to thermal collapse than A + T- rich molecules. When DNA molecules are heated above physiological temperatures (to near 100°C), their hydrogen bonds break and the complementary strands often separate from one another (DNA denaturation). Because the G...C base pair is held together by 3 hydrogen bonds, and A...T base pair by only 2 hydrogen bonds, the higher temperatures are necessary to separate G + C-rich strands than to break apart A + T-rich molecules.

In recent years many evidences have been collected in favour of the Watson and Crick model of DNA structure. They include (i) further and more x-ray diffraction studies, (ii) direct examination and precise measurements of the DNA molecule under the electron microscope, (iii) physico-chemical data on the so-called "melting point" of DNA, which is the temperature at which the hydrogen bonds break causing the stiff DNA molecules to collapse, (iv) enzymatic digestion studies in which the kinetics observed follow a pattern predicted from a double-stranded structure, and (v) enzymatic synthesis of DNA in the test tube by *Kornberg* (1962).

Duplication of DNA

According to *Watson* (1970), DNA carries all the specificity involved in its *self-replication* and it is important for the linear DNA molecules to be able to circularise while replicating. Initiation of replication begins at a fixed point on the circular chromosome, with growth of many chains starting by elongation of pre-existing strands. However, several enzymes are involved in replication. These are endonuclease, exonuclease, DNA polymerase and polynucleotide-ligase. Enzymes which break the internal phosphodiester bonds are called *endonucleases*. Thus, according to *Watson* (1970), replication must start with an endonuclease cut. *Exonuclease* is the enzyme that digests DNA for 3' or 5' ends of strands. *DNA polymerase* catalyses the formation of DNA from deoxyribonucleoside triphosphates, using DNA as a template. The *polynucleotide-ligase* covalently joins the backbone chains of DNA. Some of these enzymes probably also function in the repair of damaged DNA molecules. Thus, replication is an enzymatic process, and several enzymes participate in it.

The replication of DNA has been explained by the following hypotheses:

(a) *Template Hypothesis:* This is the hypothesis of *Watson* and *Crick* and it can be explained in their own words " Now, our model for DNA is, in effect, *a pair of templates*, each of which is complementary to the other. We imagine that prior to duplication the hydrogen bonds are broken, and the two chains unwind or separate. Each chain then acts as a template for the formation on to itself of a new companion chain, so that eventually we have *two pairs* of chains, whereas we only had one before. Moreover the sequence of the pairs of bases will have been duplicated exactly. This is the so called *template hypothesis* for DNA replication, and it could account for accurate and specific gene duplication as well as occasional modification of the DNA. It can be imagined that on rare occasions a nucleotide pair containing the wrong base is somehow incorporated into the newly forming complementary polynucleotide strand, thus, resulting in a mutation."

(b) *Hypothesis of Levinthal and Crane: A* precise geometrical model whereby the duplication of DNA molecule could occur in a continuous manner without requiring the prior separation of strands was suggested by C. *Levinthal* and *H.R.Crane* (1956). According to this hypothesis, the unwinding and rewinding of the DNA can occur simultaneously. They suggested that duplication could begin at one end, separating the strands, and the free energy released in forming the polymer would provide the energy for the rotation of the two lengthening daughter strands as well as the shortening parent strand.

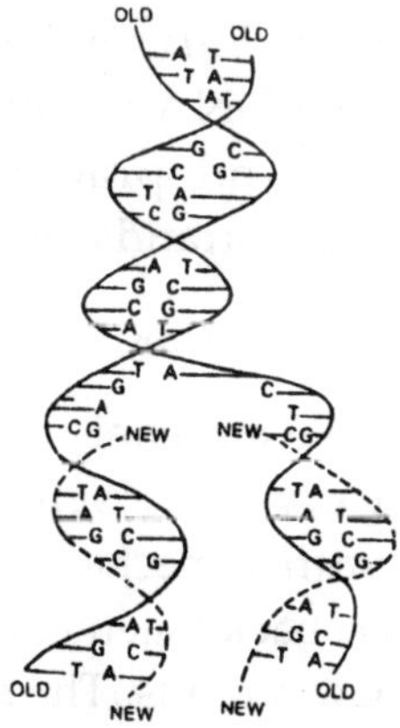

The Replication of DNA According to the Template Hypothesis of Watson and Crick.

(c) *Hypothesis of Kornberg*: According to the hypothesis of *Kornberg* (1962), as suggested in the template hypothesis of *Watson* and *Crick*, prior to duplication two strands of DNA separate by opening of hydrogen bonds. Then two new strands, having a complementary base composition with respect to the parent DNA strands are synthesised.

However, in this mechanism of the cellular duplication of DNA, which takes place in a medium of nucleotide pool, the two strands do not completely separate before the synthesis of the new strands. Instead, duplication goes hand in hand with strand separation. The monomeric precursors for DNA synthesis are the *deoxyribonucleoside triphosphates*, which are enzymatically joined together in the presence of the DNA template.

$$\underbrace{\begin{matrix}\text{dATP}\\+\\\text{dTTP}\\+\\\text{dGTP}\\+\\\text{dCTP}\end{matrix}}_{\text{the four deoxyribonucleoside triphosphates}}\ \xrightarrow[\textit{DNA polymerase}\ \text{(Kornberg)}]{\text{DNA template}}\ \text{DNA + diphosphate (2Pi)}$$

Selection of the correct nucleotides by hydrogen bonding to the correct template bases is a very accurate process. *Kornberg* (1969) has recently proposed what he calls a *"speculative model for helix replication, in vivo"* and he suggests that the enzyme DNA *polymerase* participates in achieving a sequential and practically simultaneous replication of both helical strands of DNA.

Mechanisms for DNA Replication

There are three possible mechanisms for the replication of DNA. These are conservative, semi-conservative, and dispersive. In the *conservative* mechanism, the strands of the parent double helix come back together again after serving as templates for the two new strands. The *dispersive* mechanism implies that the original strands break up and then somehow recombine with the new DNA strands.

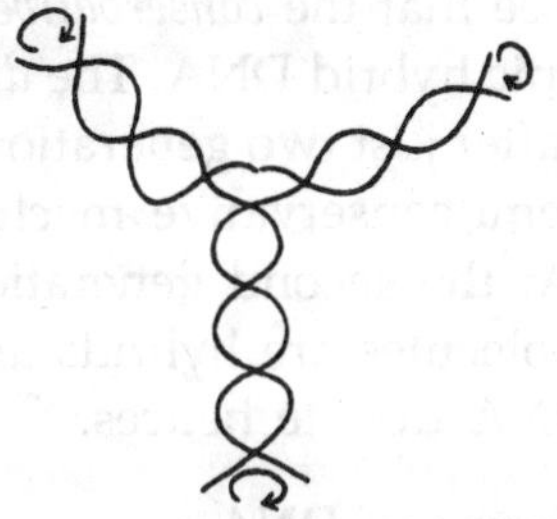

The Replication of DNA According to Levinthal and Crane.

In *semi-conservative* mechanism, half of the DNA is conserved at the molecular level. Each of the two strands of the parent molecule serves as a template for the synthesis of a new complementary strand. Thus, in the newly formed DNA molecules one strand is always new and one is always old. This mechanism has been demonstrated both at the molecular and cellular levels.

The most convincing evidence for the *semi-conservative* mechanism has been given by *M. Meselson* and *F.W. Stahl* in 1958. Working with the common colon bacillus, *Escherichia coli,* they grew some of this bacteria in a medium containing N^{15} (a heavy isotope of nitrogen) as the only source of nitrogen. As a result, all the nitrogen of the bases in the DNA of freshly grown bacteria were of the heavier N^{15}. The density of this DNA was greater than that of the ordinary DNA of this species made with N^{14}.

The analytical data were obtained in this experiment by measuring densities. Some of these bacteria grown in the N^{15} medium were then transferred to a new medium where N^{14} was the only source of nitrogen. By the time a new generation had developed, the density of the new DNA indicated that each DNA molecule must consist of one strand of "heavy" (N^{15}) DNA, and one strand of "light" (N^{14}) DNA. We may call this *"isotopic hybrid DNA"*. In the second generation, half the DNA molecules were of the isotopic hybrid type, and half were light .

Thus, we can see that the *conservative* mechanism would not yield any isotopic hybrid DNA. The *dispersive* mechanism would not, atleast after just two generations, yield any wholly light DNA. The semi-conservative mechanism is consistent with the results. At the second generation (grand daughter molecules), two molecules are hybrids and two are not, *i.e.*, they are "light" DNA double helices.

Three Structural Forms of DNA

There are three structural forms of DNA. These are: *B-form, A-form* and *Z-form,* which are described as follows:

(a) *B-form:* B-form or B-DNA exists in solutions of purified DNA. It is the naturally occurring double helical DNA existing as right-handed form.

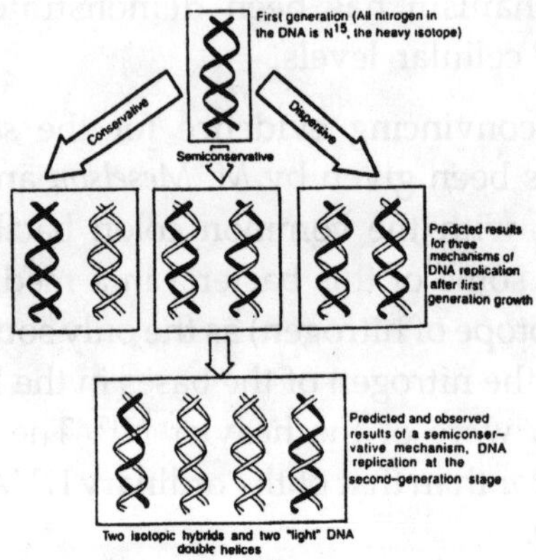

The Meselson-Stahl Experiment. Dark Strands Contain N^{15}. Light Strands Contain N^{14}.

(b) *A-form:* A-form or A-DNA is also double stranded molecule in the form of a right-handed helix. It occurs when DNA fibres are dried to about 75 per cent relative humidity. The A-form has also been detected in double-stranded RNA molecules, and in single crystals of double-stranded oligonucleotides, and the mixed DNA/RNA oligonucleotides. In A-form, the two strands are anti-parallel in their 5'-3' directions.

(c) *Z-form:* The Z-Form or Z-DNA is double-stranded in the form of a left-handed helix (*Rich*, 1979). The Z-form appears to be favoured by repeating dC - dG runs as shown below:

dC dG dC dG dC dG
dG dC dG dC dG dC

In the left-handed form, the phosphate-sugar backbone of each nucleotide follows a zig-zag pattern around the axis of the helices, hence the molecule is called Z-DNA. In the Z-helix, the conformations of the guanine and cytosine deoxyribonucleotides differ considerably from one another, so that the repeating unit becomes a dinucleotide, unlike the B-helix in which repeating unit is a single nucleotide.

While the structure of Z-DNA has been based upon studies of DNA crystals (crystals of hexanucleotide d), Z-DNA has recently been identified in polytene salivary gland chromosomes of *Drosophila.* The Z-form co-exists with the right handed form (B-DNA) in the same molecule. Now it is considered that switches in DNA helicity may be related to biological functions such as selective gene expression.

Single-stranded DNA: In a few bacterial viruses, *e.g.*, the ϕ × 174 of E. *coli*, DNA exists as a single-stranded circular DNA molecule composed of 5370 nucleotides. It contains nine genes most of which code for the structural proteins of the virus. In single-stranded DNA, the amounts of adenine and thymine, as well as guanine and cytosine, are not equal. During reproduction of these viruses, the single-stranded DNA (referred to as the "+" strand) is injected into the host bacterial cell, where it acts as a template for the reproduction of a complementary polynucleotide chain (called the "-" strand). These two polynucleotides combine to form a conventional double helix, which then serves as the template for the production of additional + strands. The newly formed + strands are then encapsulated in the viral protein coats to form new virus particles (*Kornberg*, 1979).

Satellite DNA: In human DNA, four satellite DNAs have been observed that constitute 6 per cent of the chromosomal DNA, or a total of 174,000 kb per genome. In *Drosophila melanogaster*, satellite DNAs have been separated from main-band DNA (buoyant density of 1.701g/cm^3.) by CsCl density-gradient centrifugation *(Peacock et al,* 1973). This is most readily accomplished by shearing the DNA into double-stranded segments of sizes from 20 to 60 kb before centrifugation. Under these conditions, the DNA distributes into a main band and a set of smaller bands called *satellite bands* having densities of 1.672 and 1.688 g /cm^3.

The predominant satellite DNAs generally contain very simple sequences that are repeated many times. For example, in the satellite DNA with a buoyant density of 1.672 g / cm^3 the sequence

— A C A A A C T—
— T G T T T GA —

is repeated 12 × 10^6 times. The satellite DNAs are generally found in the centromeric region of chromosomes and may act in pairing and segregation.

Biological Function of DNA in Living Beings

DNA is the primary genetic material of the cell. The 20 × 10^{-12} gram of DNA present in the human fertilized egg has the ability to determine all the hereditary characteristics of the mature human individual. The main functions of DNA are as follows:

1. *Replication:* DNA, as the primary genetic material, has the ability to replicate. Formation of new molecules of daughter DNA is called the replication of parent DNA. The replication of DNA must occur at every cell division to ensure the equal distribution of DNA in daughter cells. Thus, the most important function of DNA (gene) is the control of its own duplication.

2. *Transcription or Synthesis of RNA:* All the three types of RNA (messenger RNA, ribosomal RNA, and transfer RNA), that in turn take part in protein synthesis, are synthesised under the control of DNA by the process known as transcription.

3. *Translation or Synthesis of Proteins and Enzymes:* Translation is the guiding of amino acid sequence in proteins. Thus, DNA controls protein synthesis via RNA. Due to this ability of DNA, specific proteins and enzymes are synthesised by the cell. The relation of DNA to RNA and protein is usually summarised by the following formula, which is called the "central dogma" of molecular biology.

$$\text{DNA} \xrightarrow{\textit{transcription}} \text{RNA} \xrightarrow{\textit{translation}} \text{Protein}$$

4. *DNA and Mutation:* The genetic material or DNA possesses properties, which determine the development of many different phenotypes. All types of mutations are almost certainly related to changes in the structure of the genetic material. From what we have learnt of the Watson-Crick theory, a change in the nucleotide sequence of a DNA molecule must necessarily affect the amino acid sequence of an enzyme. Of the important physical agencies for altering DNA, atomic radiations and X-ray produce the most serious changes.

5. *DNA and Viruses:* When a *virus* (a chemical aggregate made of protein and nucleic acid) encounters its host cell, *its nucleic acid core penetrates the cell.* Once inside the living cell, the viral nucleic acid directs the making of more of itself and of protein overcoats for the new cores. When enough new particles are made, the host cell bursts, and neighbouring host cells are invaded by the cores of the newly made virus particles. Thus, the viral infection (*e.g.*, the tobacco mosaic virus infecting the leaves of the tobacco plant) spreads. One virus, that has been extensively studied, is labelled $f \times 174$, and

has *E. coli* as its host cell. The nucleic acid in $f \times 174$, which is DNA, has 5500 nucleotide units and exists as a single- stranded, closed ring. This DNA core material can be isolated from the virus and yet retain the natural infectivity of the virus.

6. *DNA as the Thread of Life:* However, the biological significance of the DNA can be judged from the statement: "DNA is the secret of life", or *"Life is the secret of DNA."* It has been called "the thread of life" by *Kendrew* (1967).

2

Role of RNA

Ribonucleic acid or RNA is the predominant nucleic acid in the cytoplasm, although all cytoplasmic RNA is of nuclear origin. In the nucleus it is present in the nucleolus, nucleoplasm and chromatin, and in the cytoplasm RNA is the main component of the ribosomes. In some viruses, *e.g.*, the tobacco mosaic virus (TMV), the RNA is present in the centre, surrounded by a protective coat of protein and serving as the genetic component of the virus. Almost all RNA molecules have a linear shape. However, recently only one circular replicating form of RNA of the mouse encephalomyocarditis (EMC) virus has been reported by *Agol et al.* (1970).

There is a great variation in the size of the linear RNA molecules. Some of them contain only a small number of nucleotides (about 75), while others have over 10,000. Independent of their length, they generally do not have a completely random shape. Some parts of the RNA molecules are double helical, while others are non-helical. In the double helical regions, the bases are hydrogen bonded to each other using DNA-type base pairing.

Chemical Composition of RNA

RNA is composed of:

(i) *A Pentose Sugar:* called ribose . It has a pentagonal ring with 5 carbons, two of which (3' and 5') are linked to phosphoric acid and a third one (carbon 1') to the base.

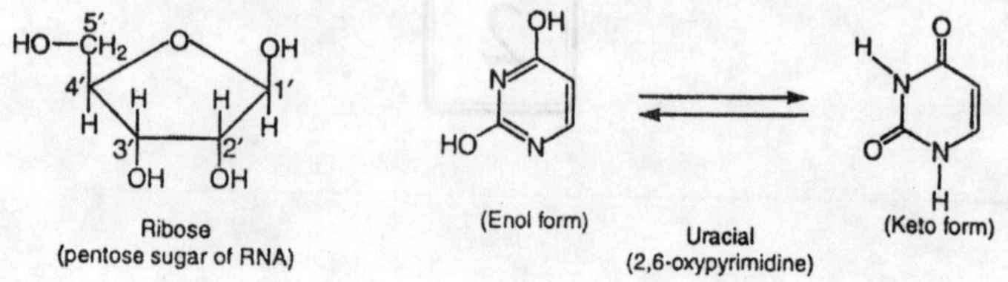

The chemical structure of Ribose and Uracil.

(ii) *Nitrogen Bases:* Usually there are four heterocyclic bases in RNA. Two of the bases are *purines.* These are adenine (A) and guanine (G). *Pyrimidine bases* of the RNA are cytosine (C) and uracil (U). There is no thymine (T) in RNA. The uracil of RNA can pair with adenine. In addition to these four usual bases, some RNA molecules (*e.g.*, tRNA) may have some " unusual " bases.

(iii) *Phosphoric Acid:* or H_3PO_4 links the nucleotides by joining the riboses of two consecutive nucleotides with an ester-phosphate bond. This bond links carbon 3' in one nucleotide with carbon 5' in the next. Linkage between a purine base and the sugar (ribose) occurs between group 9 (base) and group 1' (sugar), while pyrimidines link at group 3. The resulting combinations of the base and sugar are called *nucleosides* and are designated as: adenosine, guanosine, cytidine and uridine.

When these ribosides combine with H_3PO_4, the nucleotides of the RNA are formed. Thus, like the DNA, the RNA is also made up of nucleotides which are called *ribonucleotides* . These are as follows:

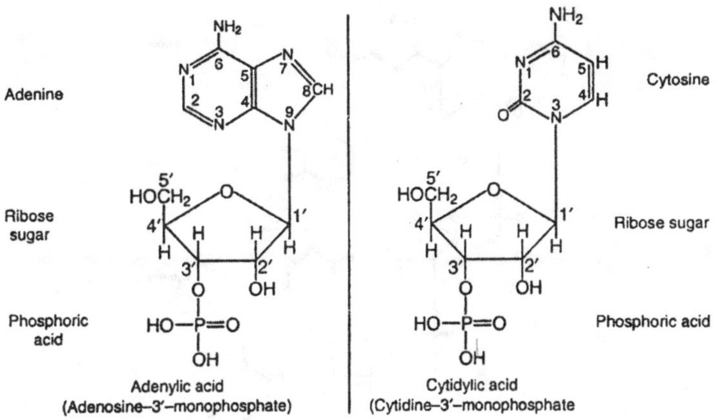

The chemical structure of two ribonucleotides.

(a) Adenosine-3'-monophosphate(Adenylic acid)

(b) Guanosine-3'-monophosphate(Guanylic acid)

(c) Cytosine-3-monophosphate(Cytidylic acid)

(d) Uridine-3'-monophosphate (Uridylic acid)

The above 3' nucleotides also occur as 5' nucleotides. In RNA polynucleotide, these four types of nucleotides are linked together by 3'-5' *phosphodiester bonds.*

Structure of Ribonucleic Acid

Like the DNA, RNA is also a long unbranched molecule consisting of four types of nucleotides linked together by 3'-5' phosphodiester bonds But there are several important differences between DNA and RNA (Table). The first is that the sugar of DNA is deoxyribose, whereas RNA contains *ribose.* The latter is identical to deoxyribose except for the presence of an additional OH (hydroxyl group). The second difference is that RNA contains no thymine but instead contains *uracil,* and the third difference is that RNA is not a regular double helix.

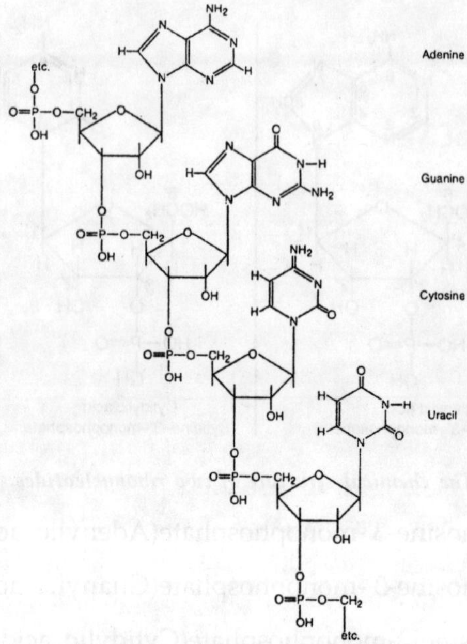

A small segment of RNA chain formed by 3′-5′ linkage of ribonucleotides.

In RNA, the amount of adenine is not equal to the amount of uracil. Similarly, the amount of guanine and cytosine also usually differ from each other This suggests that RNA does not possess a regular hydrogen-bonded structure. But unlike double-helical DNA, RNA is usually single stranded . However, the RNA strand may be twisted around itself to produce DNA-like helices at several places. In the regions of helix formation, the bases are hydrogen-bonded, *i.e.*, adenine pairing with uracil, and guanine with cytosine. Non-helical regions of RNA contain many non-bonded bases. However, the significance of non-bonded bases is not known.

The RNA-Type and their Functions

On the basis of functions and physical properties there are several kinds or classes of RNA. The principal types of RNA are: (1) Viral RNA (normally wrapped in a protein coat),

(2) ribosomal RNA, (3) messenger RNA, (4) transfer RNA, and (5) small nuclear RNA.

Significance of Viral RNA

In most plant viruses and a number of bacterial and animal viruses the hereditary material is single-strand RNA. But two groups of animal viruses called *revoviruses* contain double-strand RNA. The RNA of tobacco mosaic virus (TMV) is one of the most extensively studied and best-known molecules of RNA, which has been used for the study of virus assembly, protein synthesis, the mechanism of mutagenesis, and infectivity. Each TMV virus particle of molecular weight 4×10^7 contains a single strand of RNA of molecular weight 2.2×10^6. There is less stacking of bases in TMV RNA than in DNA. The presence of *intramolecular RNA loops* is an important part of the assembly of TMV from the RNA and protein sub-units.

Function: It was first thought that viral RNA was structurally incapable of forming a double helix. But now it is well known that in different RNA viruses a variety of mechanisms are used for viral RNA replication. They include various ways of using the RNA genome directly, and the unusual mechanism of transcribing RNA into DNA with the help of an enzyme called *reverse transcriptase* and forming double-strand DNA. This observation created a sensation when it was first discovered as it modified the "central dogma" of molecular biology from

$$\text{DNA} \longrightarrow \text{RNA} \longrightarrow \text{protein, to}$$

$$\text{DNA} \rightleftharpoons \text{RNA} \longrightarrow \text{protein}$$

The RNA to DNA arrow is shown dashed because reverse information flow from RNA to DNA does not occur very frequently.

The use of RNA, in RNA viruses, for the storage of genetic

information may represent an earlier evolutionary stage of information transfer.

The self-replicating RNA viruses lack protein coats. These *viroids* are surprisingly small; the one that causes spindle tuber disease of potatoes is among the largest and has only 359 nucleotides.

They are capable of coding only for approximately 100 amino acids, far too few for the minimum number and weight of enzymes necessary for RNA replication, but not too few for a short peptide that could alter the activity of host cell enzymes. Alternatively, the RNA itself may affect the synthesis or processing of normal cellular RNAs.

Similarities and Differences between DNA and RNA

DNA	*RNA*
1. DNA occurs mainly in the nucleus as the chromosomal DNA, but also in mitochondria and other self-reproducing organelles.	1. RNA occurs in the cytoplasm (as the main component of ribosomes), nucleolus, and chromosomes.
2. The acid component of DNA is the phosphoric acid.	2. Same
3. The pentose sugar of DNA is *deoxy-ribose,* hence the name DNA.	3. The pentose sugar of RNA is *ribose,* hence the name RNA
4. The purine bases of DNA are *adenine* and *guanine.*	4. Same
5. The pyrimidine bases of DNA are *cytosine* and *thymine.*	5. The pyrimidine bases of RNA are *cytosine* and *uracil.*
6. DNA contains only the usual bases (A, G, C and T).	6. In addition to the four usual bases (A, G, C and U), some RNA molecules may have some 'unusual' bases.
7. In DNA, adenine pairs with thymine by hydrogen bonds, and cytosine with guanine.	7. In RNA, adenine pairs with uracil by hydrogen bonds, and cytosine with guanine.
8. DNA is usually a *double helix* so that the ratios of A-T and G-C are always 1-1 to satisfy the base-pairing rules.	8. RNA is usually single-stranded having both helical and non-helical regions. Hence amount of adenine is not equal to the amount of uracil, and the amount of guanine and cytosine also differ from each other.
9. The hydrolysing enzyme of DNA is *deoxyribonuclease.*	9. The hydrolysing enzyme of RNA is *ribonuclease.*

Contd...

DNA	*RNA*
10. Cellular DNA regulates its own synthesis, *i.e.*, DNA is a self-replicating molecule.	10.Cellular RNA is not a self-replicating molecule,
11. DNA is synthesised in the nucleus in the presence of DNA template, four deoxyribonucleoside triphosphates and the enzyme DNA *polymerase*.	11. RNA is synthesised in the nucleus upon a DNA template in the presence of the four ribonucleoside triphosphates and the enzyme *RNA polymerase*.
12. DNA is Feulgen-positive, the deoxyribose is responsible for the reaction.	12.RNA is Feulgen-negative. It is detected by basophilic dyes with ribonuclease treatment.
13. DNA is the primary genetic material of the cell.	13. RNA is not the primary genetic material of the cell. It is the genetic component of some viruses (e.g., TMV) only.
14. DNA regulates the synthesis of protein via RNA because all the three types of RNA are coded on the DNA template.	14. All the three types of RNA play their role in protein synthesis,

Ribosomal RNA (rRNA): About 75 to 80 per cent of all the RNA of a cell is found in the ribosomal particles. Each sub-unit of the ribosome contains single-strand RNA . The large subunit (50 S) of the *E. coli* ribosome (70 S ribosome) has 23 S ribosomal RNA and 5 S ribosomal RNA, which have a molecular weight of about 1.2×10^6 and 4×10^4 respectively. The small subunit (30 S) of the E. *coli.* ribosome has 16 S ribosomal RNA whose molecular weight is about 0.6×10^6.

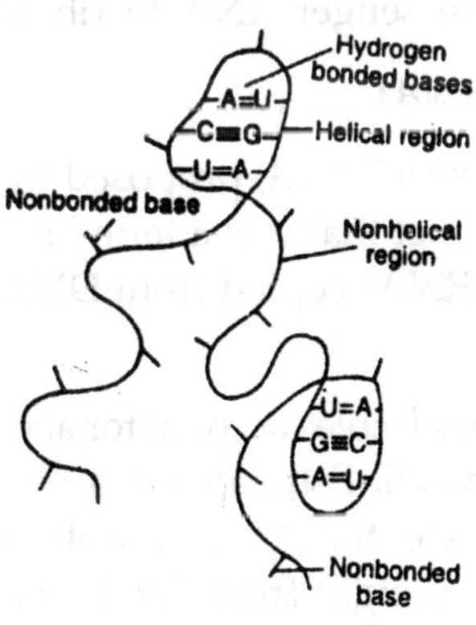

In 80 S ribosomes (with 60 S and 40 S sub-units), there are 28 S and 18 S ribosomal RNAs, with a molecular weight of 1.3×10^6 and 6×10^5 respectively. The length of a large unfolded RNA molecule has been estimated to be about 7000 Å by *Hart* (1965).

In both sub-units, the RNA molecules are associated with protein whose chief function is to help bring the template RNA and the amino acid-transfer RNA precursors together correctly. Because ribosomes easily attach basic dyes, it is thought that RNA is exposed at the surface of the subunit, and the protein may be in the interior in relation to non-helical parts of the RNA. About 60 per cent of the ribosomal RNA is helical or double stranded and contains paired bases. The helical part of the ribosomal RNA is supposed to be on the surface of the sub-units.

The base composition of the ribosomal RNAs of different sources varies considerably and it does not follow a rule of A = U and G = C. The most abundant bases are adenine and guanine. However, ribosomal RNA also contains some methyl groups.

Function: The function of ribosomal RNA is not known and at present it is difficult to say why ribosomes contain rRNA as well as protein. It is believed that unpaired bases in ribosomal RNA are in some way involved in the binding of transfer RNA and messenger RNA to ribosomes.

Messenger RNA (mRNA)

The name *"messenger" RNA*, proposed by *Jacob* and *Monod* in 1961, refers to the fact that it is a template molecule (hence also called template RNA) copied from DNA and has a rapid turnover.

Messenger RNA molecules account for about 5 to 10 per cent of the total RNA in a cell. Estimates of a molecular weight of mRNA so far indicate that it is a polydisperse material ranging in molecular weight from 5×10^4 to 5×10^5 or more.

Another characteristic of mRNA is that it is *heterogeneous.* The size of the molecule varies considerably since it is adapted to the size of the polypeptide chain for which it will code. In E. *coli,* the average size of mRNA cistron is 900 to 1500 nucleotides, thus corresponding to peptide chains consisting of 300 to 500 amino acids. Sometimes messenger RNAs are much longer if several adjacent cistrons are copied at the same time.

These are the so-called *polygenic* or *polycistronic* messengers. For example, the 10 specific enzymes involved in the metabolism of histidine may be coded in the same mRNA molecule. In higher cells, although messenger RNAs of very high molecular weight are found, some mRNAs may be *monocistronic.* For example, haemoglobin is coded by two (α and β) structural genes.

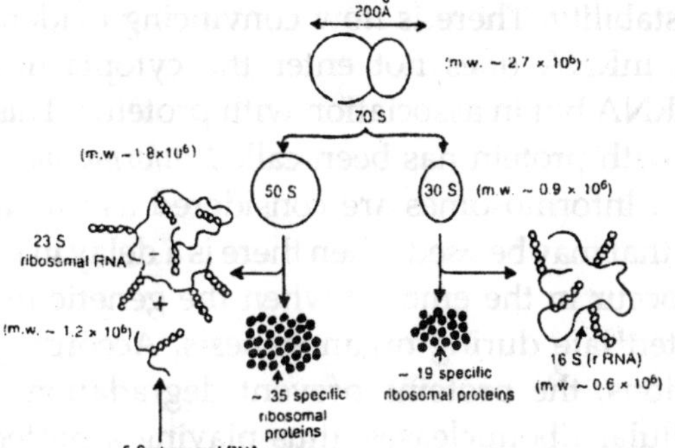

The structure of E. coli ribosome and their RNA.

Messenger RNA molecules are synthesised in the nucleus by DNA in the presence of the enzyme *RNA polymerase* and the triphosphate forms of the nucleotides (ATP, UTP, CTP and GTP) by a process that is known as *transcription.* The sequence of bases in mRNA is determined by complementary sequence on DNA. Thus, A in DNA is replaced by U in messenger RNA, C is replaced by G, G is replaced by C, and T is replaced by A. However, it appears that only either of the two strands of the DNA is used in RNA synthesis. The reason why only one strand of the double-stranded DNA is a *"coding strand"* is that mRNA otherwise would always be formed in complementary pairs and enzymes also would be synthesised in matching pairs.

Function: Messenger RNA molecules are the bearers *of genetic code* after they have been synthesised under the direct control of DNA. Once made, mRNA molecules move out of the nucleus into the cytoplasm where they attach ribosomes to themselves at intervals along their chains. Such an assembly of many ribosomes along an mRNA chain is called a *polysome* or *polyribosome.* When protein synthesis takes place, the mRNA cannot dissociate from the ribosomes. In this way, the most important property of mRNA is that it functions as a template in protein synthesis.

In eukaryotic cells mRNA molecules may have a high degree of stability. There is now convincing evidence that in these cells mRNA does not enter the cytoplasm as naked strands of RNA but in association with proteins. This complex of mRNA with protein has been called *informosome* by *Spirin et al.* (1965). Informosomes are considered as mRNA-protein complexes that may be used when there is a delay in translation. This may occur in the embryo when the genetic information is manifested late during organogenesis. According to other interpretations, the proteins prevent degradation of mRNA by intracellular ribonucleases, thus playing a protective role. It has been suggested that the protein may also be required, for recognition of the ribosomes, because when mRNA is purified, it tends to lose its template properties. The protein may also reversibly mask the mRNA and in this way control the synthesis at the level of translation. This may result in a regulation or modulation of protein synthesis.

Transfer RNA (tRNA) or Soluble RNA (sRNA): Because this RNA has a much lower molecular weight than rRNA or mRNA, it is more soluble in the media used to isolate the higher-formula-weight form. Thus, the designation "soluble" refers to a physical property and says nothing about the function. When its function is to be described, various authors refer to sRNA as *transfer RNA* (tRNA) or as *amino acid adapter RNA* (aaRNA).

In *E. coli,* tRNA accounts for 10 to 15 per cent of the total RNA of the organism. Each molecule of tRNA consists of a single chain of about 70 to 80 nucleotides that is folded back on itself.

The first nucleotide sequence of a specific tRNA molecule (from yeast) was worked out by *Robert Holley* in 1964-65, who suggested a cloverleaf diagram for the molecule. For this he earned the Nobel Prize in 1968. This tRNA specifically attaches to *alanine* and therefore is called *alanine-tRNA.* It contains 77 nucleotides arranged in a particular manner , and has a molecular weight of about 25,000.

The most striking feature of the alanine -tRNA is the presence of several "unusual" bases, i.e., the bases other than adenine, guanine, cytosine, or uracil. These unusual bases differ from the normal bases by the presence of one or more methyl (CH_3) groups. Most of the methyl groups are enzymatically added after the nucleotides are linked together by 3'-5' phosphodiester bonds. However, the function of the unusual bases is not clear. They cannot form the usual base pairs and may thus have the function of disrupting the double -helical regions, thereby exposing the free keto and amino groups which can then form secondary bonds to mRNA, ribosome, or to the enzyme required to attach a specific amino acid to its specific tRNA molecule.

The important aspects of the suggested *cloverleaf model,* which shows the two-dimensional folding pattern, for tRNA molecules are as follows.

The 3' end of the molecule consists of ACC (adenylic acid, cytidylic acid, cytidylic acid) sequence and a variable fourth nucleotide. The amino acid always attaches to the 3' terminal A. The terminal nucleotide of the 5' end is usually guanylic acid. On moving along the backbone away from the 3' end we find the acceptor stem, T stem, and the T loop or the first loop of the cloverleaf. The T loop is so named because it includes thymine (T). It is made up of 7 unpaired bases and always contains the sequence 5'-T pseudo UCG-3', *i.e.*, towards the 5' end the first base is *ribothymidine* and it is followed towards the 3' end by pseudouridine, cytosine, and guanine, etc. The first loop binds to the ribosomal surface during protein synthesis.

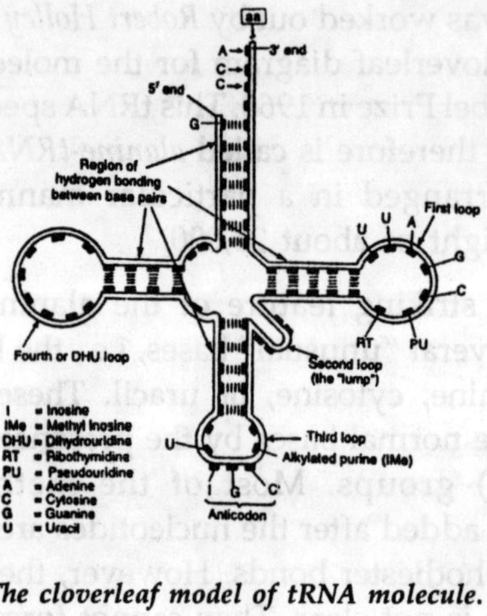

The cloverleaf model of tRNA molecule.

The first loop is followed by a loop of highly variable size often called "the lump" or the extra arm or the variable loop, which in different tRNAs has different numbers of nucleotides ranging from 4 to 21. The third loop or the anti-codon loop consists of 7 unpaired bases, of which three form the *anti-codon* and bind to the mRNA codon during protein synthesis. The anti-codon is bracketed on its 3' end by an alkylated purine base (methyl inosine) and on its 5' end by uracil. The fourth non-hydrogen-bonded section of the molecule or the fourth loop consists of 8 to 12 unpaired bases and is relatively rich in *dihydrouridine*. Hence, it is called DHU loop or D loop, and may be involved in binding to the specific activating enzyme (amino-acyl synthetase).

Besides the above non-hydrogen-bonded sections the alanine-tRNA molecule consists of four regions of hydrogen bonding between the base pairs. Thus, a majority of the bases are hydrogen-bonded to each other because the number of adenine is equal to the number of uracil, and similarly, the number of guanine is equal to the number of cytosine.

Thus, in a given tRNA molecule we may recognise four specific sites. These are: (a) the amino acid attachment site, (b) the codon recognition site (anticodon), (c) the ribosome recognition site, and (d) the specific site for recognising the amino acid-activating enzyme.

An interesting feature of the cloverleaf diagram is the presence of nucleotide sequences that are constant in all the 100 of the tRNA sequences that have been determined so far. The number of base pairs on the stem regions is also constant; 7 in the acceptor stem, 5 in the T stem, 5 in the anti-codon stem, and 3 or 4 in the D stem. These features are maintained in tRNA molecules from bacteria, plants and animals. Indeed, the pattern of loops, stems and constant nucleotides found in the tRNA cloverleaf appears to have the same universality as the genetic code.

In 1974, the three-dimensional structure of the yeast phenylalanine-tRNA has been described by *Alexander Rich* and *Sung Hou Kim* from an X-ray crystallographic study at a resolution of three angstroms. This L-shaped map confirmed the earlier finding that the molecule is organised into two columns of nucleotide bases stacked at right angles to each other. These columns have both helical and non-helical regions corresponding to the stems and loops of the cloverleaf model. The map further revealed the two helical regions, each consisting of about 10 base pairs corresponding to one turn of the double helix, and possessing the same type of hydrogen bonding between complementary nucleotide bases as found in the double helix of DNA.

Function: The function of tRNA is to attach to itself a particular amino acid for which it is coded and to carry it to the site of protein synthesis (specially an mRNA site in contact with a ribosome), and there give it up to the growing end of a polypeptide chain at the particular moment called for by the genetic code. For each amino acid, there is a specific activating enzyme which activates the carboxyl group of that amino acid for covalent bonding with the 3' terminal adenosine of the tRNA. The first step of the reaction uses the energy of ATP.

$$\underset{\text{(amino acid)}}{AA} + ATP \xrightarrow{\textit{amino acyl synthetase}} \underset{\text{(amino acid adenylate)}}{AA\text{-}AMP} + P\text{-}P$$

The AA-AMP so formed remains bound to the enzyme until the proper tRNA arrives to form the AA-tRNA complex.

$$AA - AMP + tRNA \xrightarrow{\textit{amino acyl synthetase}} AA\text{-}tRNA + AMP$$

For this purpose the enzyme must have two different sites : one to recognise the amino acid and another to recognise the specific tRNA. Every cell needs at least 20 different kinds of activating enzymes and at least 20 kinds of tRNA molecules. There must be at least one of each for every amino acid. It was formerly known that only 20 different tRNA molecules were used, but now it is clear that there may be more than one tRNA molecule for the same amino acid. The amino acid which is used first of all in protein synthesis is formyl methionine and the tRNA which attaches to it is called F-met-tRNA.

Thus, tRNA acts as a vehicle for transferring amino acids into the assembled chain of the protein. In other words, it plays a key role in translating the genetic code into the sequence of amino acids in a protein. Without the tRNA molecule genetic information could not be expressed in the synthesis of protein. *Alexander Rich* and *Sung Hou Kim* (1978) have stated that tRNA participates in a variety of other processes also, some of which are of great importance. For example, tRNA molecules can donate amino acids to preformed protein molecules or to the molecular structure of the cell wall in bacteria independently of the ribosome.

Another process in which tRNA participates is the control of gene expression. Certain tRNAs with an amino acid attached to them are known to determine whether genes (segments of DNA) will be expressed by regulating their transcription into messenger RNA or not. However, the details of the mechanism are not known, but in some systems the control function is associated with a particular modified nucleotide in the tRNA molecule, say, a uracil that has been converted into a pseudouracil. It is thought that tRNA helps to control the expression of many different genes, although the exact number is not known. Some other observations suggest that tRNA may be involved in still more types of biochemical regulation. For example, in embryonic development one kind of modification of certain nucleotides in a tRNA gives way to another kind. Similarly, when a normal cell becomes cancerous, the kinds of modifications of nucleotides in its tRNA molecules change substantially. However, it is not yet known whether these transformations are associated with the regulatory functions of tRNA.

Now it is known that tRNA is associated not only with the synthesis of polypeptide chains but also with that of polynucleotide chains. This synthesis is carried out by special enzymes such as *reverse transcriptase.* The latter, discovered a few years ago as a constituent of several rumour viruses, synthesises a strand of DNA from a template of single-strand RNA, a direction of information flow that is the reverse of the normal one. For this a specific type of tRNA first binds to the enzyme and to the viral RNA and signals the synthesis of the DNA copy to begin. But why a tRNA serves this purpose is not known at present.

Small, Nuclear RNA and its Characteristics

The small, nuclear RNA (snRNA) is a metabolically stable RNA molecule found in the nucleus of the cell. It occurs predominantly in ribonucleoprotein particles. It is a small molecule consisting of 100 to 300 bases. The RNA found in the cytoplasm of eucaryotic cells is synthesised in the nucleus. However, much of this RNA never leaves the nucleus and thus never participates directly in protein synthesis.

RNA Polymerases and their Applications

In *E. Coli,* the protein that catalyses the synthesis of RNA using DNA as a template, the *RNA polymerase* is structurally well known. The holoenzyme is composed of five non-covalently bonded polypeptides two α sub-units with MW 40,000; P with MW 150,000; β ′ with MW 160,000-and α (sigma) subunit with MW 90,000.

In eucaryotic cells, three RNA polymerases are found, each having great subunit complexity. They can be distinguished from one another by their sensitivity to the fungal toxin α -amanitin and by their cellular location. RNA polymerase 1 (also known as A) occurs in the nucleolus, is resistant to α -amanitin, and transcribes the large ribosomal RNAs. RNA polymerase 2 (also known as B) occurs in the nucleoplasm, is sensitive to low concentrations of α -amanitin, and is the main enzyme responsible for synthesising the heterogeneous nuclear RNA that later becomes processed into mRNA. RNA polymerase 3 (also known as C) is also in the nucleoplasm, is sensitive only to high concentrations of α -amanitin, and synthesises tRNAs and 5S rRNA. There are also RNA polymerases found in mitochondria and chloroplasts.

3

Establishment of Masses

Somatic Embryogenesis (SE) is a tissue culture approach where proliferative embryo suspensor masses are established from non-meristematic cells and subsequently cultured to produce organised bipolar structures possessing shoot and root meristems (that is, somatic embryos). The term somatic refers to embryos developing asexually from vegetative (or somatic) tissue. This method has been used-in horticulture, agriculture, and to some extent, in forestry-as a means of rapidly multiplying elite varieties or clones. Through the application of bulk-handling techniques, SE is cost-effective as a propagation approach for conifers.

There have been major advances over the past 25 years in the development of operational vegetative propagation systems for conifer species used in plantation forestry programmes. These propagation systems provide a means of bringing new genetic material into forestry programmes through the capture of a greater proportion of the gain from additive and non-additive genetic components inherent within a selected tree species. Vegetative propagation systems also provide a method for multiplying superior families identified in tree improvement programmes. Vegetative propagation systems being developed for use in forestry utilise the following approaches:

1. Rooted cuttings
2. Micropropagation through organogenesis tissue culture
3. Somatic embryogenesis

Currently, rooted cuttings are the most effective propagation technique that is operationally available to multiply specific individuals that have desirable traits. In a recent survey, over 65 million conifer cuttings are produced annually around the world, with this number growing rapidly. The primary use of rooted cutting technology is for bulk production of genetically improved materials. Production of rooted cuttings is essentially a 2-step process: the production of cutting-donor plants and the production of rooted cuttings.

Donor plants can range from selected trees of wild populations to trees grown from seeds genetically improved families under an intensive nursery cultural regime. Rooted cuttings of forest tree species are most successfully produced from juvenile portions of donor plants because these juvenile portions of a plant will provide cuttings with the potential for good initiation of root primordia. Cuttings are placed in a rooting environment (that is, high humidity and soil moisture, warm soils, and moderate light levels), allowed to develop roots, then treated as rising 1-year-old seedlings.

Organogenesis is a tissue culture system that relies on the multiplication of shoots or the de novo formation of organs originating from either unorganised callus or preformed shoots or induced buds. Propagules produced through this system are essentially treated as microcuttings. Thus, shoot propagules are placed in an optimal rooting environment and treated in a similar manner as cuttings. Distinct from conventional cuttings, SE offers the advantage of long-term storage of germplasm through cryopreservation. Although considerable research focus and efforts have demonstrated the potential for

SE propagation of a large variety of plant species, success has been achieved only at the research scale.

Research and development in the area of SE of commercial conifers has been driven by the following 2 factors:

— The multiplication of superior families
— The selection of elite clones from among such families to capture a greater proportion of the genetic gain

In conifers, embryogenic cultures can be induced from developing and/or mature zygotic embryos or young germinants, but not from older explants. Thus, SE fulfils a role similar to cuttings with respect to the multiplication of families. The value-added traits that can be captured and propagated through SE are those that can be identified through any tree-breeding programme and include yield, wood quality, plus stress, pest and disease resistance.

Spruce species appear to be most amenable to SE propagation technology. Interior spruce- a complex of white and Engelmann species Picea glauca (Moench) Voss x Picea engelmannii Parry ex Engelm is a major commercial species in British Columbia, Canada. Within British Columbia since the 1980 approximately 100 million container interior spruce seedlings have been planted annually in forest regeneration programmes. To supply improved genetic material for these programmes, a tree improvement programme has been developed for large areas of the commercial land base. This breeding programme affords the opportunity to match significant potential genetic gains with the technical opportunity to deliver value-added material to operational reforestation programmes through SE technology.

Protocols for Somatic Embryogenesis

In general, the SE process is divided into several phases:

— Culture initiation-induction
— Proliferation-maintenance, multiplication, suspension culture

— Cryopreservation - germplasm storage
— Maturation - somatic embryo production
— Embryo drying
— In vitro germination and early growth

All parts of the laboratory process are performed under-sterile conditions to prevent microbial contamination. Because technology is most advanced for Picea spp., the methodologies are those developed for interior spruce. Apart from the stage of commercial development, the most significant protocol difference among the various conifer species is at the initiation phase. For spruce, stored seed can be used for initiation of SE, whereas for many other conifers, most notably Pinus spp. and Douglas-fir, fresh immature cones are required, with initiation success restricted to a 1- to 2-week developmental window (Cyr unpublished data).

Initiation: For spruce, mature zygotic embryos are dissected from the seed and placed onto semi-solid medium containing plant growth regulators. The dissected embryo is referred to as the primary explant. The induction of SE is first evidenced by the growth of new tissue, usually at the hypocotyl-cotyledon junction of the explant. This process takes approximately 8 to 12 weeks and the new tissue is referred to as "putative" embryogenic tissue or embryonal suspensor mass.

Proliferation: The embryonal suspensor mass is characterised by the presence of early-stage somatic embryo structures that are analogous to those occurring during normal seed development. This tissue is isolated from the primary explant and transferred to fresh medium. Under these conditions; the tissue multiplies and develops as early-stage somatic embryos. Tissue is subdivided (subcultured) and transferred to fresh medium every 7 to 10 days. This material consists of multiple embryos and is often referred to as embryogenic callus. Sufficient material can be produced for cryopreservation within 4 to 6 weeks.

Cryopreservation: Cryopreservation facilitates long-term storage of the valuable germplasm produced by SE.

The embryogenic tissue is treated with cryoprotectants, frozen to -35°C under controlled conditions, and then subsequently stored in liquid nitrogen (-196°C). Cryopreserved tissue can be regenerated within 1 to 2 weeks after a simple thawing process.

Maturation/Embryo Drying: To advance the development of somatic embryos, tissue is transferred to a medium containing the phytohormone abscisic acid. Within a period of 4 to 7 weeks, this results in the production of mature somatic embryos that are analogous to their zygotic counterparts.

To complete the process and effect a transition to germination upon exposure to suitable conditions, somatic embryos are harvested and subjected to a drying process. This step enhances the vigour of the somatic embryo. At this stage, the embryos are often referred to as synthetic seed, or "synseed."

Germination/Early Growth: Spruce synthetic seed does not require stratification. The embryos are placed on germination media containing no phytohormones and supplemented with a carbohydrate. In vitro germination occurs within 5 to 7 days and proceeds to the development of true needles at 4 to 6 weeks. At this point the plantlets can be transferred to ex vitro conditions.

Throughout the 1990's, germinants from SE technology have shown continued improvement in their development into high-quality somatic seedlings in the nursery. In addition, somatic seedling propagation technology has been successfully integrated into the container seedling production system.

This containerised delivery system accounts for upwards of 95 per cent of the conifer seedlings grown for reforestation programmes in British Columbia, Canada. Through this time period, the performance of somatic seedlings has been compared to zygotic seedlings.

Production of conifer seedlings in container nursery systems is primarily dictated by 2 operational criteria:

— Maximise greenhouse production through the greatest number of plantable seedlings per unit area

— Minimise the number of seedlings that do not meet defined morphological standards at the end of the production cycle

The goal of the interior spruce somatic seedling programme is to produce seedlings that meet these normal operational criteria and have a high level of seedling quality. For every nursery production cycle over the past 5 years, survival of somatic seedlings after 1 growing season has consistently averaged 95 per cent. This indicates that germinants are of a high quality coming out of the laboratory. This high survival count is important because it demonstrates that a single sowing of these germinants into container cavities is an effective use of both somatic seedlings and greenhouse production space. How the germinants initially respond to the nursery environment will have a profound influence on subsequent morphological development.

During the first stages of development in the nursery, past studies have found variable success in obtaining normal shoot growth of somatic, compared to zygotic, seedlings. As a result of this variable performance, initial production runs of somatic seedlings resulted in a higher than desired culling rate (up to 20 per cent) based on British Columbia Ministry of Forests shoot-height culling standards. Recent nursery performance studies of somatic seedlings have shown that a proper nursery cultural environment (that is, nutrients, temperature, and moisture) during the initial establishment stage will result in normal morphological development of seedlings.

Shoot growth of zygotic and somatic seedlings was similar during first-year development for seedlings grown during the 1995 nursery production cycle.

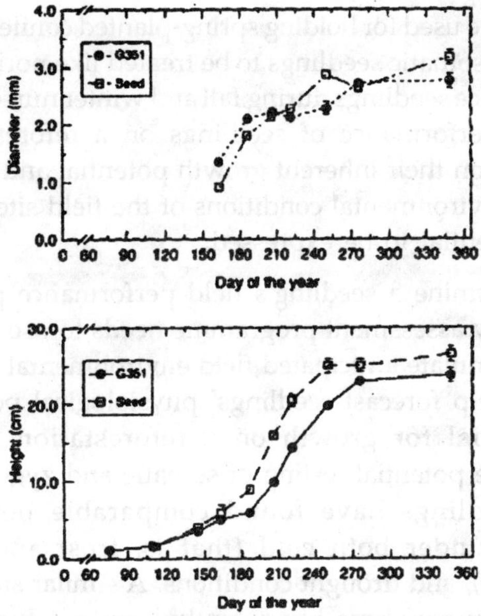

Somatic (G351) and zygotic interior spruce seedling height and diameter growth (mean ± SE) of 1 + 0 container stock from one family during the nursery production cycle.

Due to this improved morphological development, culling rates for 1996 have decreased to 5 to 8 per cent. Thus, currently produced somatic seedlings consistently meet morphological standards required for production of operational containerised interior spruce seedlings.

Under normal nursery production procedures for interior spruce, seedlings need to develop the proper level of winter hardiness before they are lifted for storage of the 1+0 frozenstored stocktype or placed outside of the greenhouse into outdoor holding compounds for the second year of a container production cycle for the 2+0 summer-plant stocktype.

During fall acclimatisation, somatic and zygotic seedlings have similar dormancy, freezing tolerance, and root growth patterns. In addition, somatic and zygotic seedlings have similar performance throughout a normal 5-month period of frozen

(-2°C) storage used for holding spring-planted conifer seedlings. This allows somatic seedlings to be treated like normal zygotic interior spruce seedlings during fall and winter nursery cultural practices. Performance of seedlings on a reforestation site depends upon their inherent growth potential and the degree to which environmental conditions of the field site allow this growth potential to be expressed.

To determine a seedling's field performance potential, a stock-quality assessment programme needs to use an array of tests that simulate anticipated field environmental conditions. This will help forecast seedlings' physiological performance and potential for growth on a reforestation site. Field performance potential testing of somatic and zygotic interior spruce seedlings have found comparable performance capability under both cold (that is, frost and low soil temperature), and drought conditions. A similar stock quality assessment programme has recently been conducted on the 1996 somatic seedling production run and results are comparable to previous findings. This indicates a consistency in the quality of somatic seedlings to be planted on reforestation sites.

Reforestation site trials have tested the field performance of interior spruce somatic seedlings in comparison to zygotic seedlings. These trials have found that somatic and zygotic seedlings have comparable summer seasonal water relation patterns and gas exchange response patterns. In addition, somatic seedlings have comparable or better performance than zygotic seedlings in response to damaging winter conditions. Somatic and zygotic seedlings have corresponding incremental height and diameter growth over at least 2 seasons in the field, resulting in similar new root and shoot development.

The survival rates of somatic and zygotic seedlings after 2 growing seasons are comparable, 83 and 77 *per cent*, respectively. Thus, performance of somatic seedling performance over 2 growing seasons on a reforestation site indicates that they have all of the traits that are desired in container stock for use in forest regeneration programmes.

Tree Improvement Integration Programme

There is an opportunity with SE technology to not only capture additional gain in height but also take advantage of family differences and potential clonal differences in the resistance to white pine weevil- Pissodes strobi Peck-inherent in interior spruce. This clonal strategy involves an "add-on" testing phase to the main breeding programme. The breeding, testing, and selecting of parents for additive gene effects is the main emphasis of the tree improvement programme for interior spruce.

First-generation clonal seed orchards have been rogued, and genetic gains for improved shoot growth from this orchard were about 11 per cent based on 10-year height results. Clonal testing of progeny from first-generation selected parents will capture additional gain for improved shoot growth over and above gains from rogued first-generation seed orchards. About 1,300 clones have been produced for field trials from 31 full-sib and 14 open-pollinated weevil-resistant families of interior spruce using SE technology.

Embryogenic callus from all of these clones have been cryopreserved for long-term storage until field selections are made. Parents of the full-sib families were selected, as judged by their progeny, based on 15-year height and 10-year weevil resistance. The open-pollinated families were selected on the basis of provenance trials. Ramets derived from SE clones are being field-planted to assess for accelerated growth and increased levels of resistance to weevil attacks. Ultimately, a total of 30 to 50 clones will be selected based on mean height growth and weevil damage starting 5 to 6 years after establishment of the trials. The selected clones will be removed from cryostorage and produced as somatic seedlings that will then be deployed operationally to reforestation sites as diverse genetic mixtures.

Gains expected from the clonal programme will be 22 per cent in height, as well as the deployment of clones with minimal susceptibility to weevil attack. In parallel with this programme, morphological development and physiological performance from a subset of selected clones is ongoing to define parameters that can be used to profile selected clones. Studies have been conducted to assess the clonal variation in morphological development in the nursery and over the first 2 field growing seasons. In the nursery, there is marked variability in the morphological development that occurs between clones within the same family in response to identical nursery culture conditions.

This variability in nursery performance will have to be addressed in relation to the current concept of producing morphologically uniform stock for reforestation programmes in British Columbia. Studies have also examined clonal differences in physiological performance at selected times, and within selected environments, throughout the yearly seasonal cycle. These studies have also found that there is a wide range in physiological performance between clones of interior spruce in relation to potential field site conditions. This information will provide a means of defining clones having desired physiological traits that can tolerate limiting field site conditions.

The practical application of this programme is to develop a series of measurement parameters that can be used to assist in the profiling of clonal material that has been selected through standard genetic selection field trials and will be deployed within operational reforestation programmes.

Operational SE Production System

The ultimate test for commercial acceptance of a novel technology such as somatic embryogenesis is the ability to develop and implement a successful operational use for the technology. The key components that must be addressed during this stage are

— Development of a cost-effective manufacturing process
— Delivery of high-quality products that provide predictable and reproducible performance
— Technology validation and promotion in the market place

Plant tissue culture processes tend to require significant hands-on manipulation to optimise the quality of plant propagules produced, but the labour costs involved render the products excessively expensive and therefore unacceptable in the marketplace.

Attempts to minimise labour costs and improve product quality resulted in the focusing of research priorities during the past decade on automating some or all of the stages required for synseed production and subsequent handling of synseeds and germinated somatic seedlings. However, Cervelli and Senaratna noted that in spite of all these efforts, there were no successful commercial applications of SE to that date for either angiosperms or gymnosperms. During the past 18 months, significant progress has been made towards a reliable, high-volume cost-effective SE production system for spruce.

The details of the spruce SE process used at the laboratory-scale have been described. Rather than trying to develop and optimise automated systems for spruce SE, the major emphasis during this production scale-up period was to overlay the established fundamental principles of manufacturing production planning and control on the laboratory-scale SE process. The initial steps required extensive and precise identification and characterisation of the inputs, the action steps and related resources and time requirements, and the outputs of each step in the spruce SE process. In addition, computerised inventory tracking programmes were developed and incorporated.

The information generated provided a clear identification and understanding of the constraints within the process, and enabled numerous modifications and improvements in SE production planning, scheduling and processes. Furthermore, "standard operating procedures" for each stage of spruce SE production were developed and then strictly followed during every production run. The net effect was that SE productivity increased by more than 300 per cent without any additional labour inputs.

These product development has been assessing the compatibility of various commercial horticulture practices with SE, and then use this information to modify SE production practices to improve quality of somatic embryos at the germination stage and somatic seedlings at the nursery production stage. This market-driven approach-that is, awareness and adoption of existing proven technologies-has also resulted in the elimination of problems in the nursery development (establishment and growth vigour) of somatic seedlings.

Tree Improvement Technology in Reforestation Programmes: The use of tree improvement practices to enhance the genetic characteristics of planted seedlings is a forestry practice that consistently shows a high return on investment by increasing yields obtained from planted forests. The use of improved seed is an effective way of bringing genetic improvement to forest regeneration programmes. Seed orchards are currently used to produce seeds in large commercial quantities from trees having desired genetic traits.

However, improved seed does not provide a method to multiply specific individuals that have desirable traits. Vegetative propagation techniques provide the best means for multiplying the improved genetic resource developed from tree improvement programmes. Two criteria are considered important for the successful implementation of vegetative propagation systems within an operational forestry programme. First, the propagation system must have the ability to preserve superior candidate clones, without genetic change or further maturation, while genetic selection programmes are taking place.

The ability to propagate elite clones will require a capacity to maintain individuals in a form capable of regenerating after a minimum period of 5 to 10 years that is required in order to test and select clones in the field. Second, the propagation system has to be able to multiply selected clones in large enough numbers at a reasonable cost. If these 2 criteria can be reasonably met, the selected vegetative propagation systems can be implemented within an operational forestry programme. To have a successful rooted cutting production programme, donor material must be maintained in a juvenile state to ensure optimum rooted cutting performance.

The phenomenon of maturation in conifers is a natural developmental process; this maturation process is a major impediment in maintaining hedge orchards of elite genetic material for continual production of genetically improved rooted cuttings. Attempts to propagate mature individuals of northern conifer species by this propagation method has been unsuccessful or severely limited by plagiotrophism or other effects of maturation. An alternative practice for rooted-cutting production currently in use to produce juvenile cuttings is seedling-origin hedges.

Vegetative propagation of conifer seedlings from organogenesis on an operational scale has also been conducted on a limited basis. The major limiting factor in applying this propagation method in forestry has been the cost relative to the low value of individual propagules. In addition, most coniferous species have proven fairly recalcitrant to standard micropropagation methods and problems, usually resulting from early maturation of the plants that are produced.

An operational organogenesis programme on radiata pine is currently being conducted by Tasman Forestry Ltd. in

New Zealand. This programme produces just over 1 million propagules, using elite genetic material selected from their seed orchards, on an annual basis for their high-yield plantation forests. Currently, no more than 20 per cent of their annual plantings are clonal, with genetic diversity broadened by selecting only 2 to 3 clones/non-related family within the mixed family groups of 15 to 20 non-related families. If there are currently operational limitations to this system, they are the labour-intensive procedures required in handling cultures to produce quality propagules and the limited ability to store (that is, cold-storage for up to 6 years) while field selections are being conducted.

The same problem of maturation that occurs in rooted cutting programmes can also occur in organogenesis tissue culture programmes. Thus, the higher per-unit propagule cost limits the use of this technology to propagate only the very elite genetic material derived from a tree improvement programme, whereas limited storability of cultures reduces the time available to make proper genetic selections. Somatic embryogenesis is the only vegetative propagation technology that provides long-term preservation of the selected genetic component of a conifer species. Embryogenic cultures can be proliferated in a juvenile form for long periods of time to produce unlimited numbers of propagules from the same clone. These cultures can also be frozen-stored indefinitely while genetic selection is being conducted.

Somatic embryogenesis protocols have shown no evidence of somaclonal variation, thus ensuring the genetic stability of spruce embryogenic clones. This propagation technology also provides a means of incorporating, through genetic engineering, selected traits that will enhance seedling performance, for example, insect resistance. Thus, SE is a vegetative propagation technology that will allow for the preservation of superior genetic material that can be used within an operational regeneration programme. Once selections of clones having desired genetic traits are made, these clones need to be brought forward to produce propagules that will then be grown under standard nursery cultural practices for the production of somatic seedlings.

Somatic embryogenesis as a viable operational propagation technology for conifer species is just coming of age. Forestry companies and government organisation around the world are currently working at bringing this technology to a point where it can produce conifer somatic seedlings, on a cost effective basis, with desired genetic characteristics-for example, Douglas-fir, loblolly pine, and radiata pine. Currently the programme at the Forest Biotechnology Centre and Silvagen Inc., BCRI, is the' furthest along in the scale-up process of this propagation technology. Somatic embryogenesis of interior spruce has developed to the point where an operational production programme has been initiated.

The emphasis of this programme in the coming years will be to scale up this propagation technology while reducing the cost per production unit. At the same time, the overall mandate will be to improve the quality of somatic seedlings from a range of genetic sources having desired field performance characteristics. This will allow for the successful implementation of SE technology as a viable vegetative propagation system within operational forestry programmes.

4

Multiplication of Plants

Plant propagation is the multiplication of plants by both sexual and asexual means. The study of propagating can provide a lifetime of challenges and opportunities to learn more about this fascinating craft, or a basic knowledge of it can provide the home gardener with the skills and techniques to keep their garden well stocked with new plants.

Methods of Propagation

There are many methods of propagating plants as there are reasons for wanting to do so, but there are basically two types of propagation — sexual and asexual. Nearly all plants in nature have the ability to reproduce sexually, that is, by seed. Along with producing seed they have developed many modifications that aid in the dispersal of that seed. Such modifications include:

- Seeds being enclose deposit the seeds elsewhere.
- Seeds with wing-like or tufted appendages that enable the seed to be carried by the wind. Seeds that are hooked or barbed that are easily attached to the coats of animals or out clothing and carried away.

— Seeds that can float thousands of miles away to wash up on a tropical island (such as a coconut).

Many types of plants in nature have also evolved means by which they can reproduce asexually. Some such means include strawberry runners, potato tubers, and Johnsongrass rhizomes. Along with all the natural modifications, people have developed many ways to propagate plants more efficiently and ways that meet the needs of both the agricultural communities and the horticultural trade.

Seeds can be treated in various ways to achieve better and more uniform germination rates. Cuttings of various plants can be rooted in greenhouses when the parent plants are under three foot of snow.

The highly specialised techniques involved in micropropagation allow growers to produce thousands of genetically identical plants, tissues (thus the term tissue culture), or cells. Learning about these procedures and many others makes plant propagation a tremendous way to expand one's knowledge of plants and gardening, and can lead to an interesting and rewarding profession.

Sexual Plant Propagation

Sexual propagation of flowering plants, as opposed to ferns, mosses, etc., begins with flowering, followed by pollination, fertilization and seed production. Seeds are used in large-scale agriculture and forestry operations for growing wheat, corn, alfalfa and tree seedlings for reforestation projects. Propagation by seeds is also critical to many aspects of horticulture including the establishment of many turfgrasses, bedding plants, and a wide range of trees and shrubs, although propagation of many of these types of plants is not restricted to sexual propagation.

Sexual propagation has several advantages when compared with asexual methods.

— It is generally the cheapest method.

— It generally requires the fewest skills, specialised equipment or facilities, and is thus the easiest method.

— Seed can be easily stored, often for several years, and still successfully germinated.

— Seeds are cheaply and easily shipped or transported around the world

— If properly cleaned and stored, seeds are less likely to carry diseases.

Seedlings are likely to be genetically different from parent plant—this may be desirable for research, breeding, plant selection and conservation work, but may be undesirable in regards to other interests. Some disadvantages of using this method include:

— Some plants don't produce live, (viable), seeds, and thus can not be grown this way.

— Seeds may take a long time to grow into mature plants.

— Seedlings are likely to be genetically different from parent plant, and may not have the same desirable characteristics.

The genetically difference is often especially true for many cultivars and hybrids.

Seeds Development: For centuries people have grown plants for the beauty of the flowers - the spectrum of colours, the multitude of shapes, the intoxicating fragrances and the meaning we have associated with various flowers in relation to holidays and traditions. However, the basic function of flowers is to be pollinated so seeds can develop, grow and perpetuate the species. It is obvious just by casual observation that all flowers are not created equal, but flower types can be grouped in terms of their structure and how it relates to pollination and seed production.

Perfect flowers are individual flowers that have both male and female parts. The male part of a flower is known as the stamen and is made up of the anther and the filament.

The female part of the flower is known as the pistil and is composed of the stigma, style and ovary. Perfect flowers may have one or many stamens and pistils in each flower. Imperfect flowers lack one or more of the parts that make up the stamen or pistil. Some flowers contain only male or only female parts. When a plant develops separate male flowers, (staminate flowers), and separate female flowers, (pistillate flowers), and both occur on the same plant, the plant is referred to as monoecious.

Examples of monoecious plants include corn, walnuts and many conifers. With corn, the tassels at the top of the plant are the male flowers, and the silks and young ear represent the female flowers. Dioecious plants have separate pistillate and staminate flowers, but they are always on separate plants, thus you will have plants with only female flowers and others with only male flowers. Dioecious plants include hollies, date palm, asparagus and Ginkgo biloba. Male plants will never produce seeds or fruit.

Females will only set fruit if a compatible male plant is nearby, neither male nor female plants will die if they are not close to each other however. Pollination involves the transfer of pollen grains from the anther to the stigma. A variety of bees, butterflies, moths and birds are responsible for pollinating a wide range of plants. Flowers often attract pollinators by various characteristics such as colour and colour patterns, shape, fragrance, offer of food, or, in the case of many orchids, resembling a potential mate for the would-be pollinator. Other plants rely on wind to carry the pollen grains.

Conifers and grasses, including many of our grain crops, are wind pollinated. Once the pollen grain lands on a receptive stigma a pollen tube begins to grow downward through the style to the ovary where fertilization occurs. Often there are many ovules (eggs) within the ovary. The ovary will develop into the fruit and each ovule will develop into a seed.

Growing Plants from Seed: The success achieved when growing plants from seed is dependent on several factors including the seed itself, the medium and the conditions to which the seed and seedlings are subjected. The seed used should be of high quality and usually, the newer the seed, the better the germination and subsequent seedling growth will be. The medium refers to the soil into or onto which the seeds are sown.

Whatever seed sowing mix is used, it should be free of weed seeds, harmful insects and pathogens that may prevent germination or kill the young seedlings. A disease known as damping-off is one of the most troublesome problems encountered if soil and sowing containers are contaminated. The mix should also be free-draining and should not remain soggy after watering. There are several ingredients that can be used to create a suitable medium.

Common components include peat moss, vermiculite (a mineral), perlite (a material of volcanic origin—often mistakenly identified as Styrofoam in soil mixes), sand, fine milled bark, compost and milled sphagnum moss. Fortunately there are several very good, bagged mixes available in garden centres, but as you experiment with growing various types of plants from seed you may prefer to mix your own special blend.

The most common bagged mixes contain peat moss and vermiculite, and often a very small amount of fertilizer to get the seedlings off to a strong start. Whatever seed mix is used, it need not be particularly rich in nutrients. It is much more important to have a mix with proper physical qualities. Once the seedlings are growing, attention can then be shifted to proper feeding with a dilute, water soluble fertilizer if necessary. Seeds can be sown into almost any type of container, but of course, clay or plastic pots or flats are logical choices.

The pots or flats should be sterilised before use (a dilute solution of bleach and water works well), and there must be drainage holes to allow excess water to drain away. When sowing the seed, the surface of the media should be even, firm (but not compacted), and slightly moistened. The depth of planting will vary with the type of plant being grown. A general rule is to plant seed at a depth of one to four times the thickness of the seed.

There are, however, some seeds that require light for germination and should be left uncovered, and just to keep you on your toes, there are others for which light can inhibit germination! This is one reason why it is wise to do a little background check on the seed you are about to sow, or at the very least, read the seed packet which often has much useful information.

Once sown, it is essential that the medium remains moist but not overly wet. Maintain the correct moisture level by placing seed pots in a plastic bag or a propagating case (creating an enclosed environment for high humidity), but keep a close eye on the temperature within. The widest range of the most commonly grown annual and biennial flowers, and vegetable plants germinate best between 65-75°F. Depending on the seeds, germination will often occur within one to three weeks, but there are others that will take longer.

Once the seedlings emerge, it is important that they receive good light, otherwise seedlings will be spindly, weak and quite difficult to transplant. It is often a good idea to lower the temperature by 5-10 of which will slow growth, but will help to keep the plants stocky. Seeds of many plants, especially woody plants and perennials, have physical or chemical inhibitors within the seed that prevent the seed from germinating even if the proper medium, temperature and moisture levels are present. The good propagator has several techniques that will effectively overcome these natural barriers to germination.

One such method is called stratification. This involves giving the seeds a moist, cool treatment. Frequently seeds are mixed with moist peat moss, vermiculite or sand, put in a plastic bag and placed in the refrigerator, (35-40°F), for a certain length of time — 90 days is a common duration. The process of soaking up water is known as imbibition, and is the first step in germination. Certain seeds have hard, impervious seed coats that prevent water from being absorbed thus delaying germination.

Scarification is a method often used to overcome this problem. Scarification involves wearing away at the seed coat to allow water in to the embryo. This can be done mechanically with a file, piece of sandpaper, or by carefully nicking the seed coat with a knife.

Some propagators dealing with large numbers of seed will use various acids to eat away at the seed coat, but this is a very refined technique in which great attention must be paid to the concentration of acid, and to the duration for which the seeds are soaked.

There are many other techniques and "tricks" that propagators employ to get certain seeds to germinate. Many plants that have adapted to habitats prone to fire have developed seeds that rely on fire as a pre-condition to germinating. Other seeds germinate only when a certain microorganism is present in the soil, or when a particular plant is already growing. With many South African plants, exposing the seed to smoke is beneficial or necessary to germination. As much as we know about seeds and how to grow them, there is still much that we do not fully understand. Until we have all the "recipes" for germinating seeds down to an exact science, a good basic knowledge of the techniques along with personal experience will go a long way in bringing success to your seed sowing efforts.

Ferns: Ferns are one of the first groups of plants to be trendy. They were all the rage in Victorian times, but have since quietly gone out of favour. That is, up until recently when there has been renewed interest in this diverse group of plants, largely from native plant enthusiasts. Ferns belong to a group of plants known as Pteridophytes which also includes mosses and others.

In terms of plant classification, they are amongst the most

primitive members of the plant kingdom. As a group, the pteridophytes are often referred to as the lower plants with flowering plants being known as higher plants. Ferns do not flower and thus do not produce seeds, but rather produce spores. Because of this, ferns have a distinctly different life cycle. Ferns typically produce a great abundance of spores which makes it possible to grow thousands of plants from a single, mature fern plant if the conditions are suitable. To propagate ferns from spores, the spores must be ripe (just as with seed).

The medium onto which spores should be sown can be the same as that used for seed sowing, although it is critical that it be sterile. Spores are dust-like, therefore sow them very sparsely, as it is very easy to sow too many! The pot containing the spores should go in a plastic bag and then be placed in a warm, bright spot, but not in direct sun. The humidity within the bag should remain high. The speed at which fern spores germinate varies greatly depending on the species. Check the bag frequently to make sure the pot doesn't dry out.

The first signs of life will be a green film developing on the surface of the medium. When this appears, mist the surface often. Gradually, young fern plants will begin to develop. When the young plants are large enough to handle easily, they can be transplanted to individual pots. Many ferns can also be propagated using various asexual methods including division, bulblets, plantlets and micro-propagation.

Asexual Plant Propagation

Asexual propagation is the production of plants using the vegetative parts of a plant. Vegetative parts include stems, leaves, roots, bulbs, corms, tubers, tuberous roots, rhizomes, and undifferentiated tissue often used in micro-propagation. Propagation by division, cuttings, layering and grafting are all forms of asexual propagation.

Although many plants can be propagated by at least one asexual method, there are some that for one reason or another can not. When compared to sexual methods, asexual methods have certain advantages.

— Plants are genetically identical to the parents so plants with desirable characteristics can be reliably cloned.

— It allows propagation of plants that do not produce seed, produce little seed, or are difficult or impossible to grow from seed.

— A grower can get a saleable or mature plant more quickly for many plants.

Some disadvantages include:

— Asexual methods are generally more expensive.

— Many asexual methods require greater skill, and/or special equipment or facilities.

— There is an increased likelihood of spreading or perpetuating certain diseases.

— Clones can become weakened and lose vigour after years of asexual production, although this is by no means a general rule.

Division of Plant: Dividing plants is probably the simplest form of asexual propagation. This method is regularly used in the propagation of a wide range of herbaceous perennials such as day lilies, Siberian iris, bee balm and ornamental grasses.

It essentially involves splitting a single large plant with many crowns or growing points into several individual smaller plants. It is labour intensive, and for that reason is a last option for commercial nurseries when there is no other viable method of propagating a particular plant.

Cuttings: Cuttings can be taken from a variety of plant parts—stems, leaves, roots, buds - but not all plants can be propagated by cuttings and certainly few, if any, can be grown from all types mentioned.

The plant from which the cuttings are taken is referred to as the stock plant or parent plant. There are many factors that effect the type of cutting used as well as the success achieved with a certain type of cutting, and include:

— The type of plant being considered for propagation.
— The age and health of the stock plant.
— The time of year.
— The facilities, equipment and material available for propagation.

When taking any type of cutting, keep in mind that by removing the cutting from the parent plant, it is cut off from its supply of moisture and is instantly under stress. This is a particularly important consideration when dealing with leafy cuttings such as herbaceous, softwood and semi-hardwood cuttings. To reduce stress on the cuttings, take cuttings on cool, cloudy days; place cuttings in a plastic bag along with a damp paper towel until they can be inserted in the rooting medium; and prepare cuttings quickly.

Also, make sure that cuttings are labelled with the plant name, the date taken and any special treatment given the cuttings, such as a particular hormone used. This can help you learn more about a plant and may be useful if you want to take cuttings again next year. The rooting medium used can vary greatly from grower to grower, and may vary depending on the type of plant being propagated. The medium must provide support for the cutting to keep it upright. It must also hold an adequate amount of moisture and allow for oxygen to reach the root zone.

Although rooting media are similar to seed sowing mixes, they are generally coarser. A mix of half peat moss and half perlite is commonly used, but many mixes exist to meet the needs of different plants, or simply produce good results for the nursery or gardener using mem. Water is the most critical aspect in the rooting process — too much and the cuttings are deprived of oxygen and the likelihood of disease is greatly increased, too little and the cuttings suffer, wilt and will root slowly, if at all. Professional growers and nurseries use mist or fog systems to maintain ideal moisture and humidity levels. These systems are controlled by a humidistat, timer, or a unit called an electronic leaf.

Light is also an important factor, at least for stem cuttings with leaves and leaf cuttings. Light is necessary for these types so the plant can continue to photosynthesise and produce carbohydrates needed for the development of roots. Too much sunlight, however is to be avoided as this can cause the cuttings to dry out too quickly. The temperature can also have an effect on root formation. Good success can be achieved with an air temperature of around 65°F for a wide range of cuttings. Often, roots will form even more quickly if bottom heat maintains the rooting media about 10°F warmer.

Auxins are one class of plant hormones that occur naturally in plants. Rooting hormones used by propagators are synthetic versions of these compounds. Used correctly, these types of hormones can hasten rooting, lead to denser root systems and help avoid certain disease problems.

They are available as liquids or powders and vary in their concentrations of active ingredient. The bases of stem cuttings are dipped into the material and then inserted in the medium. Two common rooting hormones are napthaleneacetic acid, (NAA), and indolebutyric acid, (IBA). Care should be taken when using hormones as certain cuttings can be damaged by the incorrect strength. These materials break down quite quickly in light so they should be stored in an appropriate manner.

Types of Cuttings: Stem cuttings are certainly the most important type of cuttings in regards to commercial plant production. They can be divided into four groups based on the nature and maturity of the piece of stem used—hardwood, semi-hardwood, softwood and herbaceous.

With the exception of hardwood cuttings which are often longer, cuttings of approximately three to five inches are ideal, although cuttings taken from certain dwarf plants will necessarily be shorter. When taking cuttings work with a clean, sharp, knife or hand pruners. Cuts should generally be made just below a node, (the point at which a leaf joins the stem, and the point at which roots form most readily), and the leaves on the lower 1/3 to one-half of the stem should be removed prior to insertion into the media.

The basic cutting is referred to as a simple or straight cutting. Heal cuttings are made by breaking a small, young shoot from the side of a branch. This will keep a small portion of the stem attached to the cutting. Mallet cuttings are similar but have a complete cross-section of the main stem. Some evergreens root better when heal or mallet cuttings are used. Wounding the base of cuttings of certain plants such as rhododendrons, hollies, magnolias and others can promote root production.

Wounds are generally made by stripping lower leaves and some bark from the cutting or by cutting off a thin slice of bark from the lower third of the cutting. Leaf cuttings and leaf bud cuttings are useful for propagating plants such as African violets, snakeplant, piggy-back plant and some begonias. A section of leaf, the entire leaf or the leaf and associated bud are inserted into a typical cutting media. In all cases the cutting does not became a permanent part of the plant, but gradually disintegrates alter the new, young plant is established.

A fairly wide range of plants can be propagated by root cuttings. Oriental poppies, certain species of phlox and roses, blackberries, raspberries, Japanese flowering quince among others are all likely candidates for this method. The biggest drawback to this technique is that it involves digging the parent plant out of the ground, or at the very least, severing much of the root system to get at the necessary root pieces.

Root pieces can be from 1-6 inches long depending on how coarse the roots are - the finer the roots the shorter the segments. If root cuttings are inserted into media vertically it is essential that they avoid being put in upside down, thus maintaining correct polarity. In other words, the end of the root closest to the crown of the plant should be up and the farthest point should be down. Root cuttings of some plants can be laid horizontally, sidestepping this problem altogether.

Bulbs are specialised organs with a growing point surrounded by thick fleshy scales. Tulips, onions, lilies and daffodils are all bulbous plants. Techniques for propagating bulbous plants include scaling, basal cutting, offsets and micro-propagation. Layering is yet another form of asexual propagation and is a method that encourages the development of roots on a stem while it is still attached to the parent plant.

Tip layering involves bending a branch to the ground, wounding the branch where it touches the ground and covering it with some soil. Roots develop and soon send out new shoots. At this point the new plant can be cut off from the stock plant, lifted from the ground and transplanted. Black and purple raspberries, forsythia, spirea and many other common shrubs can be grown from tip layers. Air layering is a similar technique but is a bit more involved and is used when a branch can not be bent to the ground. This technique can be used to propagate several tropical and sub-tropical trees and shrubs such as croton, rubber trees, and philodendron.

The stem is wounded and the wound is covered with a generous amount of longfibred sphagnum moss. The moss is then wrapped in a sheet of plastic that is tied off above and below the wound. The new plant can be cut off and planted once roots are visible through the plastic.

Underground Structures: Many plants produce specialised structures beneath the soil that are generally used as food storage organs on which the plant relies during adverse growing conditions. Bulbs, corms, tuberous roots, tuberous stems, tubers, rhizomes and pseudobulbs are all such organs. Frequently, though technically incorrect, all or several of these structures are referred to as "bulbs". To the propagator they are often convenient means of producing plants that have developed these sorts of structures.

A corm is made up of the swollen base of a stem surrounded by dry, scaly leaves. Crocus and gladiolus grow from corms. Corms are propagated by inducing the natural reproduction of new corms and by the cultivation of cormels, which are also naturally produced during the plants life cycle. A rhizome is a specialised stem structure that grows at or just below ground level such as in bearded Iris, lily of the valley, sugar cane and many grasses.

Typically they are easily propagated by simple division or by a special type of cutting.

A tuber is a swollen stem structure that serves as an underground storage organ with nodes, often called eyes, from which shoots emerge. The potato, Jerusalem artichoke and caladiums are all tuber-producing plants. Tubers are easily propagated by dividing them into sections, with each section containing at least one eye. Pseudobulbs, (meaning, "false bulb"), are typical storage structure of members of the orchid family. Pseudobulbs are readily separated from the parent plant as a means of propagation.

Grafting and Budding: Grafting and budding are both forms of asexual plant propagation. They both consist of connecting two pieces of living plant tissue in a way that allows the parts to unite and subsequently grow and develop as a single plant.

In any form of grafting, a piece of stem or shoot with dormant buds is the part that will grow and develop with branches. This part is known as the scion. In budding, the scion is reduced to a single bud with an attached pad of bark and cambium.

The part of the graft that will develop into the root system is known as the stock, rootstock or understock. The stock can be comprised of a root system, a sapling or, for the purposes of topworking, a mature tree that has been reduced to a trunk and main scaffold branches. Fruits and nuts, as well as roses, lilacs, dwarf conifers and many ornamentals with unique habits are examples of plants that are frequently grafted. There are several different types of grafts -splice, whip, cleft, approach, wedge, and others are all variations that have different applications for different situations and reasons for wanting to graft in the first place. The reasons for grafting are quite varied and, among others, include:

— To perpetuate clones that can not be propagated, or are not easily done so by other methods.
— To obtain the benefits of certain rootstocks, such as to control height, habit or vigour, or to impart disease resistance.
— To change the cultivar of established plants through a technique known as topworking.
— To obtain special growth habits or forms.
— To repair damaged parts of trees.

There are, however, certain disadvantages:

— It is frequently more expensive.
— Grafting and budding are fairly specialised skills, thus require great experience to be able to make grafts
— Diseases are readily transmitted
— Rootstock suckers can be troublesome and can weaken the growth of the scion.

Not just any scion and stock can be grafted successfully. The two parts must be from closely related plants. Plants from different families are incompatible and this is frequently true for plants in different genera within the same family. A scion and stock that can be successfully grafted are said to be compatible. A pairing that is incompatible will simply not grow, or will grow but never form a successful graft union leading to failure sometimes years from the time when the graft was first made. The time of year can also play a role in the success or failure of a graft. One technique that is sometimes employed to overcome the problem of incompatibility is to use an interstock.

An interstock is a piece of stem inserted between the scion and stock that forms graft unions with both. An interstock is also useful, in some cases, for imparting hardiness or a growth regulating property. When it comes to the actual act of grafting, the most critical point is that the cambium layers in the scion and stock be in close contact.

The cambium is a layer of cells between the bark and the heartwood. This layer of cells is capable of dividing and forming new cells, forming callus in the process, that are necessary in order for the graft to be successful. It is important that the cambium layers do not dry out during the grafting process. The graft union is where the scion and stock are joined.

Micropropagation

Micro-propagation, or tissue culture as it is also called, is the most cutting-edge means of propagating plants. It involves propagating plants from very small plant parts, tissues or cells in specialised conditions in which the growing environment and nutrition are strictly controlled. The basic principles of tissue culture have been known for about 100 years and such theories were suggested as long ago as the early 1800's. By 1939 scientists in the United States and France had made significant discoveries.

Within another ten years researchers had laid a solid foundation for today's large-scale tissue culture laboratories, propagation facilities, and further advances through ongoing research. It was not until recent decades that micropropagation became a feasible means of producing plants for the nursery industry. Today, many plants are propagated in this way. In catalogues, the names of plants that have been propagated by tissue culture methods are often followed by "TC" in parentheses. With every passing year more and more advances are made, and an increasingly wide range of plants find their way into the many tissue culture facilities which are appearing at an equally rapid pace.

Tissue culture makes use of an in vitro system. In vitro is from the Latin for "in glass", which is in reference to the fact that plant tissues are developed in test tubes and flasks under laboratory conditions. The multiplication of plants in vitro does not create a new process within the plant, it simply directs and enhances the plants natural potential to put forth new growth and multiply in a highly efficient and predictable way. There are several advantages to micropropagation when compared with traditional asexual methods of propagation.

— Plants can be mass-produced very rapidly.

— A new plant can generally be introduced to the nursery industry more quickly.

— Tissue cultured plants are free of insect and disease pests when removed from test tubes.

— The growth of in vitro cultures requires little care on a day to day basis, apart from casual surveillance.

There are also several disadvantages.

— More expensive. The start-up costs for a commercial micropropagation facility are high.

— The techniques used require greater skill and training.

— Not all plants can be produced through tissue culture.

— Mutations may occur during the Culturing process resulting in plants different from the parent. This can be disastrous if not noticed at an early stage.

The success of tissue culture for reproducing new plants is based on the ability of small plant parts, tissues or cells to undergo rapid cell multiplication under the proper chemical and physical conditions, and then to differentiate into the various parts that make up an entire plant. The plant part, tissue or cell type that is removed from a stock plant for purposes of being cultured is known as the explant.

The explant can be a shoot tip, root tip, leaf tissue, pollen grain, seedling tissue, bulb scales and others. There exist a number of factors that will effect the success of generating new plants by micropropagation. Sterility is of the utmost importance at all stages. Lab conditions are essential and much of the great expense is attributable to the need for such facilities. All surfaces with which the explant may come in contact, including countertops, tools and human hands, need to be sterile.

The growth medium and glassware in which the new plants will be cultured must also be sterile. Apart from sterility, the explant itself and the culture conditions (light, medium, temperature) all play significant roles in the success.

The type of explant taken from a parent plant will also effect the generation of new cell growth and the subsequent new plants. Certain explant types work better for certain plants. The medium used in tissue culture is unlike that used in any other type of plant propagation.

A semi-solid, gelatinous material called agar is used. This provides support for the culture, but by itself is essentially inert. What is mixed in with the agar is what stimulates new growth. These ingredients will vary depending on the plant being cultured, as will the concentrations used. Ingredients include inorganic salts of many essential plant nutrients and organic compounds like carbohydrates, vitamins, various hormones and growth regulators.

The four sequential stages in all tissue culture systems are:

1. Establishment
2. Multiplication
3. Pre-transplant
4. . Transplant

The purpose of the establishment stage is to establish a sterile explant in culture. The initial explants from the first stage have developed a mass of shoots that are separated into individual propagules and transferred to a fresh medium culture. This second medium is frequently the same or similar to that used in Stage 1, but the concentrations of certain ingredients may be altered. The pre-transplant stage is necessary to prepare the grown propagules, now known as plantlets, for the shift from the rigidly controlled in vitro environment to that of a more typical plant growth environment, usually in a greenhouse.

The pre-transplant stage also offers an opportunity to cull mutated propagules or ones that somehow became infected with a disease. The transplant stage is the point at which the plantlets are moved to a pasteurised soil mix essentially as seedlings would be transplanted. At this point the plantlets are very tender and dry air and/or bright sunlight can easily burn them. Gradually, humidity levels can be reduced and more sunlight provided to the young plants at which point they should be established and growing under standard conditions.

Bio-reactor Technology for Plant Propagation

Regeneration is potentiated through two morphogenic pathways: organogenesis-the formation of unipolar organs, and somatic embryogenesis-the production of bipolar structures, somatic embryos with a root and a shoot meristem. All plant somatic cells once isolated and cultured in vitro are capable of expressing totipotency. The injured cells in the outer layers of the isolated explant evolve ethylene that induces the inner layers of cells to undergo dedifferentiation.

After the loss of coordinated control, which is ensued by cell division, the formation of new gradients of endogenous phytohormones in the dedifferentiated cells enhance cell divisions in response to the various growth regulators added to the medium. Cell division can take place in an unorganised pattern with callus formation, or in an organised pattern with the formation of meristematic centres directly in the explant tissues.

Re-differentiation in the callus tissue takes place when new physiological gradients are formed in the non-organised parenchymatous tissue along with the formation of meristemoids (in vitro meristems), which can further differentiate into organised structures. In several plant species, meristematic centres form directly on the explant with very little or no callus formation and can develop into either shoots or somatic embryos. Plant regeneration can follow after the initial stages of controlled re-differentiation through either the organogenic or embryogenic pathways.

Plant regeneration in vitro is dependent on the manipulation of the inorganic and organic constituents in the medium, as well as the type of explant and the species. In most plants, successful regeneration from the callus or directly from the explants takes place after a series of subcultures in various media, in a sequence which is often , specific to the species, variety, or the newly introduced genotype.

The determining factors are the combination of the concentration in relation to medium volume and the composition of growth promoting and retarding regulators in the medium, the physiological status and competence of the cells and their capability for morphogenetic expression. Micropropagation (in vitro propagation of axillary and/or adventitious buds as well as somatic embryos) is presently used as an advanced bio-technological system for the production of identical pathogen-free plants for agriculture and forestry.

The technique, however, is still costly due to intensive hand manipulation of the various culture phases and is not used commercially for all plant species. In addition, in some plants, the initial stage of establishment and response is slow and the survival of the plants in the final stage ex vitro is often poor, which further reduces the micropropagation production potential. Efficient commercial micropropagation depends on rapid and extensive proliferation along with the use of large-scale cultures for the multiplication phase.

Furthermore, normal plant development during the acclimatisation and hardening stage is mandatory to ensure a high per cent of survival after transplanting to the greenhouse. Mechanisation and automation of the micropropagation process can greatly contribute to overcoming the limitation imposed by existing conventional labour-intensive methods. Considerable attention has been directed towards automation of the repeated cutting, separation, subculture, and transfer of buds, shoots, or plantlets during the multiplication and transplanting phases.

Progress in tissue culture automation will depend on the use of liquid cultures in bio-reactors, which will allow fast proliferation, mechanised cutting, separation, and automated dispensing. These techniques were reported for some plants and were shown to reduce hand manipulation and thus reduce in vitro plant production costs. The various propagation aspects of several plant species in bio-reactors and some of the problems associated with the operation of bio-reactors were recently reviewed by Takayama and Akita.

One major problem addressed was microbial contamination as affected by both the introduced plant material and the operation procedures of large-scale bio-reactors. Liquid media have been used for plant cells, somatic embryos, and organ cultures in both agitated flasks or various types of bio-reactors. Although the use of bio-reactors has been directed mainly for cell suspension cultures and secondary metabolites production, research directed at improving bio-reactors for somatic embryogenesis has been reported for several plant species.

The cultivation in liquid media using a temporary immersion system with different frequencies of immersion was reported to improve plant quality and multiplication rates of banana, coffee, and rubber. Bio-reactors were used also for the cultivation of hairy roots mainly as a system for secondary metabolite production. The use of a simplified acoustic window mist bio-reactor for transformed roots and for carnation shoots was reported to improve biomass growth.

Information on the use of bio-reactors as a system for plant propagation through the organogenic pathway, although limited to a small number of plant species, is presently being applied to several ornamental and some vegetable and fruit crop plants. However, liquid cultures confer several problems associated with abnormal plant development, a phenomenon described as hyperhydricity, previously termed "vitrification", which causes poor plant development in vitro and later in ex vitro.

Attempts were carried out to overcome hyperhydricity in liquid cultures by the use of growth retardants, inhibitors of gibberellin biosynthesis, which decreased hyperhydricity, reduced shoot elongation, and induced bud or meristem cluster formation. The clusters were shown to be an alternative propagation system for bio-reactor cultures, providing a biomass with limited leaf elongation. Recently micropropagation of pineapple was scaled up using clusters in a temporary immersion system (ebb and flow) in bio-reactors.

Bio-reactors are presently being used for commercial micropropagation in the USA, Japan, Taiwan, Korea, Cuba, Costa Rica, Holland, Spain, Belgium, and France for ornamental and bulbous plants, pineapple, potato, and forest trees.

The cost per propagule unit of foliage plants was estimated to be reduced from 17 to 6-7 cents (US). A recent review by Takayama and Akita describes several bio-reactor techniques using plant propagules for large-scale propagation of potato, gladiolus, lilies, strawberry, Hyacinth, Amaryllis, and several Araceae.

Somatic Embryogenesis: The use of liquid cultures for the cultivation of somatic cells in recapitulating embryogeny was reported by Steward et al. and Reinert as early as 1958 in carrot. The underlying concept for cell totipotency and the ability to renew growth and express morphogenesis was based on the assumption that isolation and bathing of the cells in a specific medium will simulate the zygotic embryo conditions in the ovary.

The bathing liquid medium that contains various nutrients and growth regulators was assumed to be a source for similar stimuli that were present in the zygotic embryos' immediate environment. In addition, the segregation of embryogenic from non-embryogenic cells amplified the expression of cell totipotency. Somatic embryogenesis was achieved by the initial use of an induction medium containing an auxin, usually 2,4-dichlorophenoxyacetic acid (2,4-D), and coconut water; the latter was mainly a source of cytokinins, inositol, and reduced nitrogen compounds.

Once the pro-embryonic clusters formed, the removal or lowering of the auxin concentration initiated a sequence of events similar to zygotic embryogeny. Somatic embryogenesis was observed at first mainly in members of the Umbelliferae cultured in liquid media and has since been reported in gymnosperms and angiosperms, including several ornamentals, and vegetable and field crops, as well as perennial woody plants. The embryogenic pathway, as opposed to the organogenic pathway, can be a more efficient and productive system for large-scale clonal propagation.

Somatic embryogenesis may result in less variation through chimerism. However, somaclonal variation will be less likely to take place in direct regeneration, but might be more pronounced in somatic embryos developing from continuously cultured callus tissue. Since the embryo contains both a root and an apical shoot meristem, the rooting stage required in conventional in vitro bud or shoot propagation technology is obviated. Somatic embryos are small and can be adequately handled in scaled-up procedures.

They are amenable to sorting and separation by image analysis, dispensing by automated systems, and can be encapsulated and either stored or planted directly, with the aid of mechanised systems. Somatic embryogenesis in liquid shake or bio-reactor cultures was reported in carrot, in caraway, poinsettia, alfalfa, celery, in Eschcholtzia californica, in Nerine, in Octea catharinensis, in sweet potato, in rubber, and in spruce.

However, the ultimate goal of production of synthetic seeds through successful encapsulation of somatic embryos from liquid shake or bio-reactor cultures was reported only for carrot, alfalfa, celery, and white spruce and is currently under continuous investigation for several other species.

Organogenic Pathway: The propagation of most plants is presently carried out commercially through the organogenic pathway in agar-gelled cultures, even though the protocols are long and costly. The advantages for the mechanisation of the process are mainly for achieving a reduction in hand manipulation and labour costs. The process has to be scaled up using liquid cultures in bio-reactors to amend it to automation.

The information on the use of bio-reactors for unipolar structures such as protocorms, buds, or shoots is limited, mainly because of the problems of hyperhydricity of the leaves and shoots. One of the first attempts in using liquid cultures for micropropagation of buds was reported for orchids that formed protocorms with minimal shoot elongation during the liquid culture stage.

The protocorms could be separated and induced to form new plants after subculture to agar-gelled medium. Several other ornamental plants also have been propagated through the organogenic pathway in liquid shake cultures and bio-reactors. By controlling shoot growth and providing culture conditions that reduced abnormal leaf growth and enhanced the formation of bud or meristematic clusters, as was shown for potato, gladiolus, and Ornithogalum dubium, a high proliferation rate was achieved without the phenomenon of hyperhydricity.

Bud or Meristem Clusters: The development of spherical meristematic or bud clusters in liquid cultures was shown to provide a highly proliferative and rapid growing system amenable to automated inoculation, control of the medium components, mechanical separation, and efficient delivery to the final stage for plant growth and development.

Cluster formation appears to be associated in most species with the continuous submergence, circulation, and agitation of the plant biomass in the medium, as well as with a balanced ratio of growth-promoting and growth-retarding regulators. Clusters can form from axillary buds and/ or adventitious buds as well as from meristemoids in pro-embryogenic callus that later differentiate to somatic embryos.

The formation of condensed organised structures in which the shoots are reduced to buds or meristematic tissue was reported in several plant species. The clusters were made up of densely packed meristematic cells, actively dividing and forming new meristematic centres on the outer surface. The meristemoids surrounded loosely packed cells in the centre and exhibited some vascularisation, as was shown in liquidcultured poplar clusters.

In banana, on the other hand, the clusters were made up of condensed buds surrounding a central core with a cavity. Halperin in 1967 described the formation of pro-embryogenic masses in carrot root explants which were induced by the addition of 2,4-D to the liquid medium, resembling orchids protocorm-like bodies that formed when isolated buds were cultured in liquid medium in shake cultures. Protocorm-like clusters also were induced in liquid-cultured gladiolus buds and in several species of the complex Brodiaea by the addition of paclobutrazol or ancymidol (gibberellin biosynthesis inhibitors) to the medium.

In chicory, nodules were induced on leaves by IBA and BA in the medium. In potato and banana, bud clusters were induced by a balanced ratio between kinetin and ancymidol in the liquid medium. In woody species, McCown et al. and Aitken-Christie et al. described nodules in poplar in liquid medium and in radiata-pine in agar-gelled medium respectively, induced by the use of a balanced ratio of growth regulators.

Levin et al., working with several ornamental species, described a severalfold increase of an organogenic biomass of clusters which proliferated in bio-reactors and were separated mechanically prior to dispensing to agar-gelled cultures for further growth. The production of clusters in Philodendron cultured in liquid medium required the presence of benzylaminopurine (BA) and an inductive treatment for 24-48 h with ancymidol.

The inductive treatment prevented a carry-over dwarfing effect of ancymidol on leaf and shoot development after transplanting the clusters to agar-gelled medium for further plant growth. Gladiolus clusters cultured in a disposable 2-L bio-reactor proliferated and made up to 60 per cent of the vessel volume after 4-5 weeks. The aggregates which reached 0.5-1.5 cm in diameter tended to sink to the bottom of the bio-reactor as the biomass increased and could not be re-suspended unless the rate of aeration was increased.

An increase from 0.5 to 1.5 vvm (volume air/volume medium per minute) provided the required aeration for recirculation of the cluster biomass. Potato and banana bud clusters proliferated in bubble and disposable plastic bio-reactors at a faster rate than on agar-solidified medium. Growth regulators, when given in a specific balanced ratio of promoting and retarding substances, apparently act as morphogenic signals and control the development of the spherical meristematic or bud clusters.

The spherical meristematic or bud clusters were described by various terms: pro-embryogenic mass (PEM) in carrot, protocorms in orchids, nodules in poplar and radiata-pine, nubbins in daylilly, and meristematic or bud clusters in gladiolus, Brodiaea, Nerine sarniensis, fern, Philodendron, potato, banana, and Ornithogalum dubium.

It is suggested that the term cluster should be used, as it is more appropriate for describing these proliferative, usually rounded structures, which develop from buds or other proliferative tissue in liquid culture, through either the organogenic or somatic embryogenesis pathways.

Anomalous Plant Morphogenesis: Scaling-up in bio-reactors requires the use of liquid instead of agar-gelled media during the proliferation and biomass production stages. The culture of plants in liquid medium is known to cause anomalous morphogenesis, resulting in plant hyperhydricity. The plants that develop in liquid media are fragile, have a glassy appearance, with succulent leaves or shoots and a poor root system.

The leaves are the organs affected most severely in liquid cultures. They develop an unorganised mesophyll tissue that is made up mainly of spongy parenchyma tissue with large intercellular spaces, a deformed vascular tissue, and an abnormal epidermis.

The epidermal tissue in hyperhydric leaves lacks a well-developed cuticle and possesses malfunctioning guard cells which cannot respond to closure signals.

Hyperhydricity affects plant survival after transplanting and causes loss of the in vitro developed leaves, or even whole plants, which often wilt and die. The two major processes carried out by the leaves, photosynthesis and transpiration, are not fully functional in hyperhydric leaves and thus cause the poor performance of the transplanted plants ex vitro. In many plants propagated in vitro, the anomalous morphology and anatomical aberration often observed in plants result from deviation from the normal course of morphogenic events that are manifested in plants in vivo.

Some of the deformation, such as hyperhydric, malformed leaves and shoots, abnormal embryos, recurrent embryogenesis, and several other disorders, are apparently the result of interruption or faulty timing of the signals involved in the normal sequence of organisational events known to exist in vivo. These are problems that are manifested more severely in liquid medium and await further studies in order to understand and to help control plant morphogenic events in bio-reactor cultures.

Plant Cell and Tissue Growth in Bio-reactors

The advantages provided by aerated liquid cultures in bio-reactors include better contact between the plant biomass and the medium, no restrictions of gas exchange, the control of the composition of both the medium and the gaseous atmosphere, and the ability to manipulate the plant biomass in relation to the medium volume. The need for efficient circulation and mixing of the plant biomass, especially for cluster and embryogenic tissue, is essential to prevent sedimentation and allow optimal growth.

Various types of bio-reactors with mechanical or gas-sparged mixing were used for plant cell cultures, to provide stirring, circulation, and aeration. Mechanically stirred bio-reactors depend on impellers, including a helical ribbon impeller, magnetic stirrers, or vibrating perforated plates. Aeration, mixing, and circulation in bubble column or airlift bio-reactors is provided by air entering the vessel from a side or bottom opening through a sparger; as the air bubbles rise, they lift the plant biomass and provide the oxygen gas required.

Mechanically stirred bio-reactors are used for effective mixing, aeration, dispersion of air bubbles, and prevention of large cell aggregates formation in cell suspension cultures. In recent years, large-scale cultivation of plant cells, embryos, or organs has made use of airlift or bubble column bio-reactors and, to a lesser extent, of mechanically stirred tank bio-reactors, due to the former lower shearing force properties. Plant cells are relatively large in size, have sensitive cell walls which are highly susceptible to shearing forces, and are easily damaged.

Furthermore, since plant cells, unlike microbial cultures, do not require high O_2, the use of bubble column or airlift bio-reactors was found to be adequate and advantageous for plant tissue cultured in liquid media. In addition, it was shown that mixing by gas sparging in bubble column or airlift bio-reactors lacking impellers or blades is far less damaging for clusters than mechanical stirring, since they were shown to have a lower shearing stress. The main advantage of airlift bio-reactors is their relatively simple construction, the lack of regions of high shear, reasonably high mass and heat transfer, and relatively high yields at low input rates.

A bubble free oxygen supply bio-reactor with silicone tubing was found suitable for embryogenic cell suspensions and provided foam-free cultures. For hairy root culture, an acoustic mist bio-reactor was found to increase root biomass significantly. For efficient large-scale cultures of both somatic embryos and organogenic plant tissue, the bio-reactor configuration and volume must be determined according to the mixing and aeration requirements of the specific plant or tissue propagated, as well as for minimising the intensity of the shear stress.

One major problem encountered in large-scale liquid cultures is contamination. Fungi, bacteria, yeast, and insects are a source of serious contaminations, causing heavy losses of plant material in commercial laboratories.

In scaled-up liquid cultures, the losses are even greater, as the source of contamination due to manipulation of the bio-reactor apparatus is dependent on the various stages of preparing and maintaining the equipment. In several laboratories, the operation area is kept sterile by a positive pressure airflow, which decreases contamination risks. In addition, constant screening of the plant tissue for contaminants and continuous indexing is another safeguard.

Physical and Chemical Factors

In order to control plant morphogenesis and biomass growth in bio-reactors, various culture conditions must be manipulated, i.e., the gaseous atmosphere, oxygen supply and CO_2 exchange, pH, minerals, carbohydrates, growth regulators, and the liquid medium rheology and cell density.

Gaseous Atmosphere: The atmosphere of the culture vessel is made up mainly of nitrogen (78 per cent), oxygen (21 per cent), and carbon dioxide (0.036 per cent). The culture vessel gas composition is influenced by the volume of the vessel and the extent of ventilation. Plants evolve CO_2 and consume O_2 during respiration, while during photosynthesis CO_2 is used and O_2 is produced. During the dark period, CO, levels were found to increase in cultures, and if photoautotrophic conditions prevail, its level decreased in the light. Ethylene, ethanol, acetaldehyde, and other hydrocarbons are additional components of the gaseous atmosphere in vitro.

Most of the effects of CO_2, O_2, and C_2H_4 on plant growth in vitro were reported for agar-gelled or cell suspension cultures. In bio-reactors, the control of the gaseous phase depends on the gas flow and can be easily manipulated to provide the required levels of O_2, CO_2, and C_2H_4. In airlift or bubble column bio-reactors, the air supplied is used for both mixing and aeration. The importance of aeration and the gaseous phase were shown in potato cultured in airlift bio-reactors.

Induction of tubers was inhibited under continuously submerged conditions. Microtubers developed only after the shoots elongated and reached the gaseous phase at the top of the bio-reactor. Enrichment with O_2 and manipulation of the hormonal and osmotic conditions had no effect and could not explain the phenomenon. A two-phase culture substituting the growth medium with a tuber induction with 9 per cent sucrose in the medium enhanced tuber formation from shoots that developed above the medium and were exposed to the gaseous phase. These results emphasise the importance of the gaseous phase in the bio-reactors for specific developmental phases.

Oxygen Level: Oxygen levels in liquid cultures depend on the presence of O_2 in the gas phase above and in the air bubbles inside the medium, as well as in the dissolved O_2 in the medium. Air is released through a sparger located at the base of the bio-reactor. The available oxygen for plant cells in liquid cultures determined by oxygen transfer coefficient (kLa values) is the part that dissolves in water. Its depletion as a function of the metabolic activity of the growing cell biomass can affect the culture yield. Plant cells have a lower metabolic rate than microbial cells and a slow doubling time and therefore require a lower O_2 supply.

In general, high aeration rates appear to reduce the biomass growth. The requirements for O_2 may vary from one species to another, and must be supplied continuously to provide adequate aeration, since it affects metabolic activity and energy supply as well as anaerobic conditions. The level of O_2 in liquid cultures in bio-reactors can be regulated by agitation or stirring and through aeration, gas flow, and air bubble size. The use of a porous irrigation tube as a sparger generated fine bubbles, high kLa values, low mechanical stress and provided a high growth rate.

Growth of poinsettia cell suspension in bio-reactors was inhibited when the level of O_2 dropped below 10 per cent. When the level of O_2 was elevated to 80 per cent, cell number increased to 4.9×10^5/ml as compared to 3.1×10^5/ml at 40 per cent O_2. Somatic embryo development in alfalfa and poinsettia suspension cultures was enhanced at 78 per cent and 60 per cent O_2 levels, respectively. High aeration rates were found to inhibit cell growth in cell suspensions cultured in airlift bio-reactors.

This result was explained to be due to an effect of "stripping" of the volatiles produced by the plant cells, which are apparently necessary for cell growth. Increasing O_2 levels from 21 per cent to 80 per cent, in bio-reactor cultures of Boston fern clusters, enhanced growth values (final FW -initial FW/ initial FW) from 0.61 to 0.92. Reducing O_2 levels to 10 per cent (v/v) affected cell differentiation in bio-reactor cultures of carrot embryogenic tissue. Under these conditions embryo production was severely inhibited.

The best production of somatic embryos from embryogenic

cultures of Eschcholtzia californica and carrot in a helical ribbon impeller bio-reactor was achieved under 20 per cent oxygen concentration. When O_2 was low (5-10 per cent), it inhibited biomass and somatic embryo production, while high O_2 (60 per cent) favoured undifferentiated biomass production. Additional studies are required to provide information on optimal dissolved O_2 requirements in large-scale liquid cultures.

CO_2 ***Effects:*** The effects of CO_2 were reported mainly for plants in agargelled cultures or for cell suspension cultures used for secondary metabolite production. The reports on the effects of CO_2 enrichment in sugar-free agargelled cultures suggest beneficial promotion of plant growth during plant acclimatisation and transplanting ex vivo. The contribution of CO_2 supply during the proliferation and multiplication stage in media supplied with sucrose in bio-reactors is debatable. It is implicit that if photoautotrophic conditions do not prevail, CO_2 enrichment beyond the 0.36 per cent in the air supply is unnecessary.

There are reports that high aeration rates rather than excessive oxygen inhibit growth and that reduced growth could be due to depletion of CO_2 or to the removal of various culture volatiles including CO_2. The requirement for CO_2 was not related to photosynthesis but to some other metabolic pathways involved in amino acid biosynthesis. In poinsettia, the growth of embryos in Erlenmeyer flasks was considerably higher than in bio-reactors and it was suggested that the growth difference was related to the differences in the gaseous atmosphere in the headspace. CO_2 enrichment in an illuminated bio-reactor culture of Brodiaea clusters did not affect biomass growth.

An increase from 0.3 per cent to 1 per cent gave a similar growth value under the two CO_2 levels and supply of 135 umols $m^{-2}.s^{-1}$ photosynthetic photon-flux. In Cyclamen persicum Mill, high CO_2 levels correlated with increased production of pro-embryogenic masses.

Ethylene: Ethylene level in the headspace in liquid cultures in flasks differs from that in continuously aerated bio-reactor cultures. Most of the reports on ethylene relate to agar-gelled or cell suspension cultures. In various plant species cultured in bio-reactors, the rate of aeration will affect the level of ethylene. High rates of aeration, which are often required at high biomass densities, can cause "stripping" of volatiles that are apparently important for some plants grown in culture.

In clusters of Brodiaea cultured in liquid medium, ethylene had no effect on growth, although its level was reduced in the presence of silver thiosulphate (an inhibitor of ethylene action). The level of ethylene was reduced from 0.38 to 0.12 ml/L in highly aerated bio-reactor cultures without affecting biomass growth, which was similar under both levels.

Mineral Nutrients Consumption: Revised MS medium or media with various partial modifications in the inorganic and organic constituents of MS are used for most plant species in agar-gelled or liquid cultures in vitro. The availability of mineral nutrients depends on the type of culture, whether agar-gelled or liquid, the type and size of the plant biomass, and the physical properties of the culture. Factors such as pH, temperature, light, aeration, the concentration of minerals, the medium volume, and the viscosity of the medium will determine the rate of absorption of the various nutritional constituents.

Plant cells growing in liquid cultures are better exposed to the medium components and the uptake and consumption are faster. Agar-gelled and liquid shaken cultures often dehydrate by water evaporation to the headspace and out of the vessel and such a water loss concentrates the medium. The content of a certain component in the medium is a product of its concentration and the medium volume. The effects of the medium volume and the initial strength on potato were discussed by Kozai et al.

In bio-reactors, in which either humidified air or condensers are used to prevent dehydration, the level of the nutrients in the medium is affected mainly by the absorption rate and by cell lysis. Differentiation and proliferation of micropropagated fern, gladiolus, and Nerine nodular clusters in bio-reactors was better on half-strength than on full-strength MS minerals (M. Ziv, unpubl.). This was also true for Lilium bulblets differentiating on bulb scales that were cultured in bio-reactors. A drop in pH to 4.5 and lower values and the subsequent increase to pH 5.5 was attributed to the initial utilisation of ammonium and to the uptake at a later stage of nitrate. In several species the depletion of NH_4^+ is the first limiting factor of biomass growth and somatic embryos development.

Increasing the concentration of NH^+ to 15 mM resulted in maximum somatic embryo production in carrot cultures in a helical ribbon impeller bio-reactor. A thorough and detailed study of nutrient uptake in Eschcholtzia californica embryonic cultures was achieved in a helical ribbon impeller bio-reactor. Following a lag phase of about 140 h, first NH_4^+ and then NO_4^-, PO_4^-, K^+ and SO_4^- uptake was observed to coincide with biomass increase and somatic embryo production. After 300 h, PO_4^- was exhausted, while after 600 h NH_4^+ was depleted and Mg^{++} and Ca^{++} composition continued to deplete after about 800 h. The K^+ and Na^+ ions were not depleted at that stage.

The composition of minerals, monitored during the culture of Begonia and Brodiaea in liquid media changed, with phosphate, ammonia, nitrate, and potassium depleting faster and prior to the depletion of Ca^{++} and Mg^{++}. In somatic embryos of spruce cultured in bio-reactors, 80 per cent of the ammonium was consumed by the growing biomass. In general, biomass growth is limited by the availability of phosphate, nitrogen, and carbohydrates and to a lesser extent by the availability of calcium, magnesium, and other ions.

Carbohydrate Supply and Utilisation: Cultured plants require a constant supply of carbohydrates as their source of energy. Sucrose and to a lesser extent glucose, fructose, or sorbitol are the most commonly used carbohydrates in vitro. In general, sucrose is removed rather rapidly from the medium and after 10-15 days the sucrose can be completely depleted or reduced to 5-10 g/L from an initial level of 30 g/L in both agar-gelled and liquid cultures. At the same time, glucose and fructose that appear in the medium due to sucrose hydrolysis increase in the presence of invertase in the culture medium, and can reach levels of 5-10 g/L. Cell suspensions of Catharanthus roseus cultured in a column airlift bio-reactor showed a lag phase of 5 days, during which there was a total hydrolysis of the sucrose to glucose and fructose.

In suspension cultures of alfalfa, sucrose also was hydrolysed during the first 5 days. Most of the sugars uptake occurs after day 5 and glucose is taken up preferentially over fructose. A higher yield of alfalfa embryos was obtained when 30 g/L maltose combined with NH_4^+ was used instead of 30 g/L sucrose that was used in combination with various nitrogen sources. In embryonic suspension cultures of celery, the addition of mannitol reduced cell lysis and enhanced somatic embryogenesis. When 40 g/L mannitol was added, a higher number of embryos was produced and the frequency of singulated normal embryos was increased.

In embryogenic cultures of Eschcholtzia californica grown in a helical ribbon impeller bio-reactor, sucrose uptake started after 100 h and the sucrose was depleted after 600-800 h in culture. In Picea species cultured in shaken flasks in a medium with 30 mM glucose, the carbohydrates depleted after 10-14 days. In mechanically stirred bio-reactors, a level of 60 mM sucrose resulted in the highest cell biomass and somatic embryo number. The effects of various carbohydrates on the growth of spruce somatic embryos revealed that the response was species dependent. The biomass of fern meristematic clusters in a bubble column bio-reactor was increased with the increase in sucrose concentrations from 7.5 g/L to 30 g/L, while higher levels caused a decrease in cluster growth.

Elevated sucrose concentrations caused a decrease also in the chlorophyll content of the clusters and leaves. Increasing sucrose concentrations from 30 g/L to 60 g/L in bio-reactor cultured gladiolus clusters decreased the biomass FW by more than 50 per cent. When 30 g/L or 60 g/L were combined with paclobutrazol, a further decrease in FW was observed. On the other hand, 60 g/L sucrose induced a higher DW increment, which could be attributed to an osmotic effect and the subsequent water status of the clusters.

Gladiolus clusters cultured in the presence of growth retardants had a higher level of starch, 845 as compared to 585 mg/g DW in the control. A higher number of bulblets were produced in bio-reactor-cultured bulb scales of Lilium at 30 g/L than at higher sucrose levels. However, larger size microbulbs were produced at 90 g/L than at 30 g/L sucrose in the medium.

pH Effects: The initial pH in most plant cell cultures ranges between 5.5-5.9. Since most media are not buffered, changes during autoclaving and during the biomass growth in culture occurs. A rapid drop in pH to 4.0-4.5 took place within 24-48 h in cell suspension, organ, and embryogenic cultures. These changes were related to an initial ammonium uptake and acidification due to cell lysis. However, the pH increased after a few days and reached a stable level around pH 5.0-5.5, which was related to the uptake of nitrates. In spruce species cultured in liquid medium, the pH levels were shown to increase to 6.5-6.8 after 14 days in culture.

In embryogenic cultures of poinsettia, various initial pH levels stabilised after 20 days in batch bio-reactor cultures, between pH 5.3-5.7. Oxygen levels affected pH changes by 0.2 pH units. In fern cultured in an airlift bubble column bio-reactor, the pH dropped to 4.2 after 24 h and subsequently increased to pH 5.4. Keeping a constant pH by titration with KOH did not affect the growth response. In alfalfa, the development of somatic embryos was affected by the pH. A higher rate of embryo production was observed at a constant pH of 5.5 than at a non-buffered medium or at lower pH levels.

Carrot cell differentiation in a controlled bio-reactor was affected by the pH; the highest rate of embryo production was observed at a pH of 4.3. However, embryo development was arrested before the embryos reached the torpedo stage and continued only at pH 5.8. The changes in carrot embryo development were associated with sugar uptake and ammonium depletion and can be attributed to enzyme and metabolic activity at an optimal pH. It appears that pH requirements are species and developmental stage dependent.

Growth Regulator Effects: The use of growth regulators in liquid cultures can be more effective in controlling the proliferation and regeneration potential than in agar-gelled medium due to the direct contact of plant cells and aggregates with the medium. The limited information available on regulators seems to indicate that similar levels of growth regulators were used in both agar-gelled and liquid cultures even though availability appears to be better in liquid medium. Somatic embryogenesis in many species was first induced in an auxincontaining medium that promoted rapid cell division.

Expression of the embryogenic potential was achieved in auxin-free media, or with low levels of auxin in the medium. Further embryo conversion and maturation into plants was promoted by the addition of abscisic acid (ABA), which was found to control abnormal embryo and plant growth. Bulblets development from scales in Lilium cultured in bio-reactors was higher in the presence of BA than kinetin. However, high BA had an inhibiting effect on further bulblet growth. High kinetin also stimulated bulblets differentiation, but inhibited further growth of the bulblets. However, the addition of 0.1 mg/L naphthalenacetic acid (NAA) was found to further enhance kinetin activity, which was, however, inhibited again in the presence of high (90 g/L) sucrose levels.

In embryogenic cultures of Nerine, auxin and cytokinins were used to induce proembryogenic clusters. Embryogenic expression was achieved, however, only after a short exposure to 2-iso-pentenyladenine (2iP) and further subculture in a growth regulator-free medium. Since one of the major problems in liquid-cultured plants is malformation of shoots, hyperhydricity, the induction of meristematic or bud clusters with arrested leaf growth was one of the solutions to hyperhydricity. The use of relatively high cytokinin levels and growth retardants which inhibit gibberellin biosynthesis was the most effective method to reduce shoot and leaf growth and promote the formation of meristematic clusters.

The information on abscisic acid is limited and was found effectual mainly in the later stages of somatic embryo development, promoting normal embryo growth and maturation in carrot and alfalfa embryogenic cultures. Growth of callus and leaf development in bulblets regenerated on bulb scales of Lilium in bio-reactor cultures was inhibited by abscisic acid, thus providing singulated propagules for easier handling and storage. Ethylene effects appear to be species dependent. Information on the effects of growth regulators in bio-reactor cultures is inadequate and further research is needed to optimise culture conditions.

Temperature Effects: The control of the temperature in the liquid medium inside the bio-reactor can be easily manipulated by a heating element in the vessel or by circulating water in an enveloping jacket outside the vessel. There is, however, limited information on the effects of temperature in bio-reactor cultures, which is usually kept constant at 25°C, with small day and night fluctuations. Temperature effects on potato tuber formation in an airlift bio-reactor were studied by Akita and Takayama. A higher number and larger-size tubers developed at 25°C than at 17°C; the lower temperature caused a decrease in tuber size. Ziv and Shemesh, working with potato internode explants in liquid cultures, found that tuber formation was best at a 16 h photoperiod and 18/15°C day and night temperature. When bulblets of Nerine were cultured in liquid medium, the FW increase was higher at 25°C than at 17°C. However, when subcultured from liquid to an agar-gelled bulb induction medium, rooting was better at 17°C.

Cell and Aggregate Density

Growth and proliferation of the biomass in bio-reactors depends on airflow supply for the aeration and mixing, and for the prevention of the plant biomass sedimentation. In many plants cultivated in bio-reactors, continuous aeration, mixing, and circulation cause shearing damage, cell wall breakdown, and accumulation of cell debris, which is made up mainly of polysaccharides. Cell debris accumulation results in foaming, adhesion of cells and aggregates to the culture vessel walls, and the development of a "crust" at the upper part of the bio-reactor vessel. This layer prevents adequate circulation, causing additional cell debris formation and a demand for higher aeration rates that intensify the clogging problem.

As the biomass increases and the cultures become viscous, higher rates of aeration are required to allow for oxygen supply and circulation. Medium viscosity and foaming were reduced by the use of half the concentration of MS minerals and by lowering the level of calcium in the medium. In bio-reactor cultures in which the plant biomass used was of an aggregate or cluster nature, the foaming and viscosity were not as severe as that observed in cell suspension cultures. Surprisingly, in cluster cultures in bio-reactors, even when the mixing was poor and the clusters sedimented, biomass growth was observed, suggesting that perhaps the continuous circulation accompanied by aeration was secondary in importance to oxygen supply.

Shearing stress and cell wall damage were greatly reduced in a Vibramix bubble-free oxygenated bio-reactor, in which silicone tubing was used for air supply. Embryogenic cells cultured in a helical ribbon impeller bio-reactor produced poor-quality embryos when the mixing speed was increased from 60 to 100 rpm. This was attributed to shearing stress and cell damage. The introduction of polyethylene glycol 6000 changed the rheology of the medium in alfalfa liquid cultures and improved somatic embryo development beyond the globular stage while it was arrested in a less viscous medium.

Disposable plastic presterilised bio-reactors, with a volume capacity of 2 and 5 L, were used for the micropropagation of various plant species through bud or meristematic clusters. The plastic bio-reactors were found to provide good circulation with reduced shearing damage and foaming.

The problems of cell damage, foaming, and culture viscosity can be better controlled by developing bio-reactors with an optimal shape suitable for micropropagation through meristematic or bud clusters. Clusters that are compact and consist mainly of small meristematic cells are apparently less shear sensitive than highly vacuolated cells in parenchymatous cell suspension cultures.

5

Assimilation of Food

The term "nutrition" refers to the interrelated steps by which a living organism assimilates food and uses it for growth and replacement of tissue. Plants use inorganic minerals for nutrition, whether grown in the field or in a container. Complex interactions involving weathering of rock minerals, decaying organic matter, animals, and microbes take place to form inorganic minerals in soil. Roots absorb mineral nutrients as ions in soil water. Many factors influence nutrient uptake for plants. Ions can be readily available to roots or could be "tied up" by other elements or the soil itself. Soil too high in pH (alkaline) or too low (acid) makes minerals unavailable to plants.

The term "fertility" refers to the inherent capacity of a soil to supply nutrients to plants in adequate amounts and in suitable proportions. Previously, plant growth was thought of in terms of soil fertility or how much fertilizer should be added to increase soil levels of mineral elements. Most fertilizers were formulated to account for deficiencies of mineral elements in the soil. The use of soilless mixes and increased research in nutrient cultures and hydroponics as well as advances in plant tissue analysis have led to a broader understanding of plant nutrition. Plant nutrition is a term that takes into account the interrelationships of mineral elements in the soil or soilless solution as well as their role in plant growth. This interrelationship involves a complex balance of mineral elements essential and beneficial for optimum plant growth.

Essential Mineral Elements

The term essential mineral element (or mineral nutrient) was proposed by Arnon and Stout. They concluded three criteria must be met for an element to be considered essential. These criteria are:

1. A plant must be unable to complete its life cycle in the absence of the mineral element.
2. The function of the element must not be replaceable by another mineral element.
3. The element must be directly involved in plant metabolism.

These criteria are important guidelines for plant nutrition but exclude beneficial mineral elements. Beneficial elements are those that can compensate for toxic effects of other elements or may replace mineral nutrients in some other less specific function such as the maintenance of osmotic pressure. The omission of beneficial nutrients in commercial production could mean that plants are not being grown to their optimum genetic potential but are merely produced at a subsistence level.

There are actually 20 mineral elements necessary or beneficial for plant growth. Carbon (C), hydrogen (H), and oxygen (O) are supplied by air and water. The six macronutrients, nitrogen (N), phosphorus (P), potassium (K), calcium (Ca), magnesium (Mg), and sulphur (S) are required by plants in large amounts. The rest of the elements are required in trace amounts (micronutrients). Essential trace elements include boron (B), chlorine (Cl), copper (Cu), iron (Fe), manganese (Mn), sodium (Na), zinc (Zn), molybdenum (Mo), and nickel (Ni). Beneficial mineral elements include silicon (Si) and cobalt (Co). The beneficial elements have not been deemed essential for all plants but may be essential for some. The distinction between beneficial and essential is often difficult in the case of some trace elements.

Cobalt for instance is essential for nitrogen fixation in legumes. It may also inhibit ethylene formation (Samimy, 1978) and extend the life of cut roses. Silicon, deposited in cell walls, has been found to improve heat and drought tolerance and increase resistance to insects and fungal infections. Silicon, acting as a beneficial element, can help compensate for toxic levels of manganese, iron, phosphorus and aluminum as well as zinc deficiency. A more holistic approach to plant nutrition would not be limited to nutrients essential to survival but would include mineral elements at levels beneficial for optimum growth. With developments in analytical chemistry and the ability to eliminate contaminants in nutrient cultures, the list of essential elements may well increase in the future.

The use of soil for greenhouse production before the 1960s was common. Today a few growers still use soil in their mixes. The bulk of production is in soilless mixes. Soilless mixes must provide support, aeration, nutrient and moisture retention just as soils do, but the addition of fertilizers or nutrients are different. Many soilless mixes have calcium, magnesium, phosphorus, sulphur, nitrogen, potassium and some micronutrients incorporated as a pre-plant fertilizer. Nitrogen and potassium still must be applied to the crop during production. Difficulty in blending a homogenous mix using pre-plant fertilizers may often result in uneven crops and possible toxic or deficient levels of nutrients.

Soilless mixes that require addition of micro and macro-nutrients applied as liquid throughout the growth of the crop, may actually give the grower more control of his crop. To achieve optimum production, the grower can adjust nutrient levels to compensate for other environmental factors during the growing season. The absorption of mineral ions is dependent on a number of factors in addition to weather conditions.

These include the cation exchange capacity or CEC and the pH or relative amount of hydrogen (H+) or hydroxyl ions (OH–) of the growing medium, and the total alkalinity of the irrigation water.

Cation Exchange Capacity (CEC): The Cation Exchange Capacity refers to the ability of the growing medium to hold exchangeable mineral elements within its structure. These cations include ammonium nitrogen, potassium, calcium, magnesium, iron, manganese, zinc and copper. Peat moss and mixes containing bark, sawdust and other organic materials all have some level of cation exchange capacity.

pH: The term pH refers to the alkalinity or acidity of a growing media water solution. This solution consists of mineral elements dissolved in ionic form in water. The reaction of this solution whether it is acid, neutral or alkaline will have a marked effect on the availability of mineral elements to plant roots. When there is a greater amount of hydrogen H+ ions the solution will be acid (<7.0). If there is more hydroxyl OH- ions the solution will be alkaline (>7.0). A balance of hydrogen to hydroxyl ions yields a pH neutral soil (=7.0). The range for most crops is 5.5 to 6.2 or slightly acidic. This creates the greatest average level for availability for all essential plant nutrients. Extreme fluctuations of higher or lower pH can cause deficiency or toxicity of nutrients.

Role of Mineral Nutrients

Role of essential and beneficial mineral nutrients that are crucial for growth are explained below. Eliminate any one of these elements, and plants will display abnormalities of growth, deficiency symptoms, or may not reproduce normally.

Macronutrients: Nitrogen is a major component of proteins, hormones, chlorophyll, vitamins and enzymes essential for plant life. Nitrogen metabolism is a major factor in stem and leaf growth (vegetative growth). Too much can delay flowering and fruiting. Deficiencies can reduce yields, cause yellowing of the leaves and stunt growth.

Phosphorus is necessary for seed germination, photosynthesis, protein formation and almost all aspects of growth and metabolism in plants. It is essential for flower and fruit formation. Low pH (<4) results in phosphate being chemically locked up in organic soils. Deficiency symptoms are purple stems and leaves; maturity and growth are retarded. Yields of fruit and flowers are poor. Pre-mature drop of fruits and flowers may often occur. Phosphorus must be applied close to the plant's roots in order for the plant to utilise it. Large applications of phosphorus without adequate levels of zinc can cause a zinc deficiency.

Potassium is necessary for formation of sugars, starches, carbohydrates, protein synthesis and cell division in roots and other parts of the plant. It helps to adjust water balance, improves stem rigidity and cold hardiness, enhances flavour and colour on fruit and vegetable crops, increases the oil content of fruits and is important for leafy crops. Deficiencies result in low yields, mottled, spotted or curled leaves, scorched or burned look to leaves.

Sulphur is a structural component of amino acids, proteins, vitamins and enzymes and is essential to produce chlorophyll. It imparts flavour to many vegetables. Deficiencies show as light green leaves. Sulphur is readily lost by leaching from soils and should be applied with a nutrient formula. Some water supplies may contain Sulphur.

Magnesium is a critical structural component of the chlorophyll molecule and is necessary for functioning of plant enzymes to produce carbohydrates, sugars and fats. It is used for fruit and nut formation and essential for germination of seeds. Deficient plants appear chlorotic, show yellowing between veins of older leaves; leaves may droop. Magnesium is leached by watering and must be supplied when feeding. It can be applied as a foliar spray to correct deficiencies.

Calcium activates enzymes, is a structural component of cell walls, influences water movement in cells and is necessary for cell growth and division. Some plants must have calcium to take up nitrogen and other minerals. Calcium is easily leached. Calcium, once deposited in plant tissue, is immobile (non-translocatable) so there must be a constant supply for growth. Deficiency causes stunting of new growth in stems, flowers and roots. Symptoms range from distorted new growth to black spots on leaves and fruit. Yellow leaf margins may also appear.

Micronutrients: Iron is necessary for many enzyme functions and as a catalyst for the synthesis of chlorophyll. It is essential for the young growing parts of plants. Deficiencies are pale leaf colour of young leaves followed by yellowing of leaves and large veins. Iron is lost by leaching and is held in the lower portions of the soil structure. Under conditions of high pH (alkaline) iron is rendered unavailable to plants. When soils are alkaline, iron may be abundant but unavailable. Applications of an acid nutrient formula containing iron chelates, held in soluble form, should correct the problem. Manganese is involved in enzyme activity for photosynthesis, respiration, and nitrogen metabolism. Deficiency in young leaves may show a network of green veins on a light green background similar to an iron deficiency. In the advanced stages the light green parts become white, and leaves are shed. Brownish, black, or grayish spots may appear next to the veins. In neutral or alkaline soils plants often show deficiency symptoms. In highly acid soils, manganese may be available to the extent that it results in toxicity.

Boron is necessary for cell wall formation, membrane integrity, calcium uptake and may aid in the translocation of sugars. Boron affects at least 16 functions in plants. These functions include flowering, pollen germination, fruiting, cell division, water relationships and the movement of hormones. Boron must be available throughout the life of the plant. It is not translocated and is easily leached from soils. Deficiencies kill terminal buds leaving a rosette effect on the plant. Leaves are thick, curled and brittle. Fruits, tubers and roots are discoloured, cracked and flecked with brown spots.

Zinc is a component of enzymes or a functional cofactor of a large number of enzymes including auxins (plant growth hormones). It is essential to carbohydrate metabolism, protein synthesis and internodal elongation (stem growth). Deficient plants have mottled leaves with irregular chlorotic areas. Zinc deficiency leads to iron deficiency causing similar symptoms. Deficiency occurs on eroded soils and is least available at a pH range of 5.5 - 7.0. Lowering the pH can render zinc more available to the point of toxicity.

Copper is concentrated in roots of plants and plays a part in nitrogen metabolism. It is a component of several enzymes and may be part of the enzyme systems that use carbohydrates and proteins. Deficiencies cause die back of the shoot tips, and terminal leaves develop brown spots. Copper is bound tightly in organic matter and may be deficient in highly organic soils. It is not readily lost from soil but may often be unavailable. Too much copper can cause toxicity.

Molybdenum is a structural component of the enzyme that reduces nitrates to ammonia. Without it, the synthesis of proteins is blocked and plant growth ceases. Root nodule (nitrogen fixing) bacteria also require it. Seeds may not form completely, and nitrogen deficiency may occur if plants are lacking molybdenum. Deficiency signs are pale green leaves with rolled or cupped margins.

Chlorine is involved in osmosis (movement of water or solutes in cells), the ionic balance necessary for plants to take up mineral elements and in photosynthesis. Deficiency symptoms include wilting, stubby roots, chlorosis (yellowing) and bronzing. Odors in some plants may be decreased. Chloride, the ionic form of chlorine used by plants, is usually found in soluble forms and is lost by leaching. Some plants may show signs of toxicity if levels are too high.

Nickel has just recently won the status as an essential trace element for plants according to the Agricultural Research Service Plant, Soil and Nutrition Laboratory in Ithaca, NY. It is required for the enzyme urease to break down urea to liberate the nitrogen into a usable form for plants. Nickel is required for iron absorption. Seeds need nickel in order to germinate. Plants grown without additional nickel will gradually reach a deficient level at about the time they mature and begin reproductive growth. If nickel is deficient plants may fail to produce viable seeds.

Sodium is involved in osmotic (water movement) and ionic balance in plants.

Cobalt is required for nitrogen fixation in legumes and in root nodules of non-legumes. The demand for cobalt is much higher for nitrogen fixation than for ammonium nutrition. Deficient levels could result in nitrogen deficiency symptoms.

Silicon is found as a component of cell walls. Plants with supplies of soluble silicon produce stronger, tougher cell walls making them a mechanical barrier to piercing and sucking insects. This significantly enhances plant heat and drought tolerance. Foliar sprays of silicon have also shown benefits reducing populations of aphids on field crops. Tests have also found that silicon can be deposited by the plants at the site of infection by fungus to combat the penetration of the cell walls by the attacking fungus. Improved leaf erectness, stem strength and prevention or depression of iron and manganese toxicity have all been noted as effects from silicon. Silicon has not been determined essential for all plants but may be beneficial for many.

Photosynthesis

Photosynthesis is the process by which plants, some bacteria, and some protistans use the energy from sunlight to produce sugar, which cellular respiration converts into ATP, the "fuel" used by all living things. The conversion of unusable sunlight energy into usable chemical energy, is associated with the actions of the green pigment chlorophyll. Most of the time, the photosynthetic process uses water and releases the oxygen that we absolutely must have to stay alive.

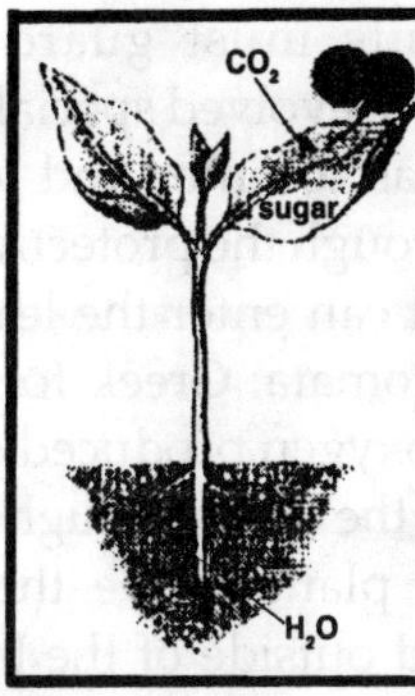

Diagram of a typical plant

We can write the overall reaction of this process as:

$$6H_2O + 6CO_2 \longrightarrow C_6H_{12}O_6 + 6O_2$$

The above chemical equation translates as: six molecules of water plus six molecules of carbon dioxide produce one molecule of sugar plus six molecules of oxygen.

Leaves: Plants are the only photosynthetic organisms to have leaves (and not all plants have leaves). A leaf may be viewed as a solar collector crammed full of photosynthetic cells. The raw materials of photosynthesis, water and carbon dioxide, enter the cells of the leaf, and the products of photosynthesis, sugar and oxygen, leave the leaf.

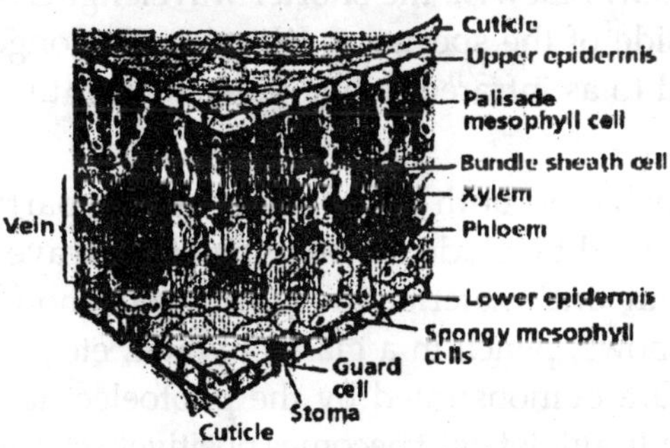

Cross-section of a leaf

Water enters the root and is transported up to the leaves through specialised plant cells known as xylem (pronounces zigh-lem). Land plants must guard against drying out (desiccation) and so have evolved specialised structures known as stomata to allow gas to enter and leave the leaf. Carbon dioxide cannot pass through the protective waxy layer covering the leaf (cuticle), but it can enter the leaf through an opening (the stoma; plural = stomata; Greek for hole) flanked by two guard cells. Likewise, oxygen produced during photosynthesis can only pass out of the leaf through the opened stomata. Unfortunately for the plant, while these gases are moving between the inside and outside of the leaf, a great deal water is also lost. Cottonwood trees, for example, will lose 100 gallons of water per hour during hot desert days. Carbon dioxide enters single-celled and aquatic autotrophs through no specialised structures.

Light: White light is separated into the different colours (=wavelengths) of light by passing it through a prism. Wavelength is defined as the distance from peak to peak (or trough to trough). The energy of is inversely proportional to the wavelength: longer wavelengths have less energy than do shorter ones.

The order of colours is determined by the wavelength of light. Visible light is one small part of the electromagnetic spectrum. The longer the wavelength of visible light, the more red the colour. Likewise the shorter wavelengths are towards the violet side of the spectrum. Wavelengths longer than red are referred to as infrared, while those shorter than violet are ultraviolet.

Light behaves both as a wave and a particle. Wave properties of light include the bending of the wave path when passing from one material (medium) into another (i.e. the prism, rainbows, pencil in a glass-of-water, etc.). The particle properties are demonstrated by the photoelectric effect. Zinc exposed to ultraviolet light becomes positively charged because light energy forces electrons from the zinc. These electrons can create an electrical current. Sodium, potassium and selenium have critical wavelengths in the visible light range. The critical wavelength is the maximum wavelength of light (visible or invisible) that creates a photoelectric effect.

Chlorophyll and Accessory Pigments: *A* pigment is any substance that absorbs light. The colour of the pigment comes from the wavelengths of light reflected (in other words, those not absorbed). Chlorophyll, the green pigment common to all photosynthetic cells, absorbs all wavelengths of visible light except green, which it reflects to be detected by our eyes. Black pigments absorb all of the wavelengths that strike them. White pigments/lighter colours reflect all or almost all of the energy striking them. Pigments have their own characteristic absorption spectra, the absorption pattern of a given pigment.

Chlorophyll is a complex molecule. Several modifications of chlorophyll occur among plants and other photosynthetic organisms. All photosynthetic organisms (plants, certain protistans, prochlorobacteria, and cyanobacteria) have chlorophyll a. Accessory pigments absorb energy that chlorophyll a does not absorb. Accessory pigments include chlorophyll b (also c, d, and e in algae and protistans), xanthophylls, and carotenoids (such as beta-carotene). Chlorophyll a absorbs its energy from the Violet-Blue and Reddish orange-Red wavelengths, and little from the intermediate (Green-Yellow-Orange) wavelengths.

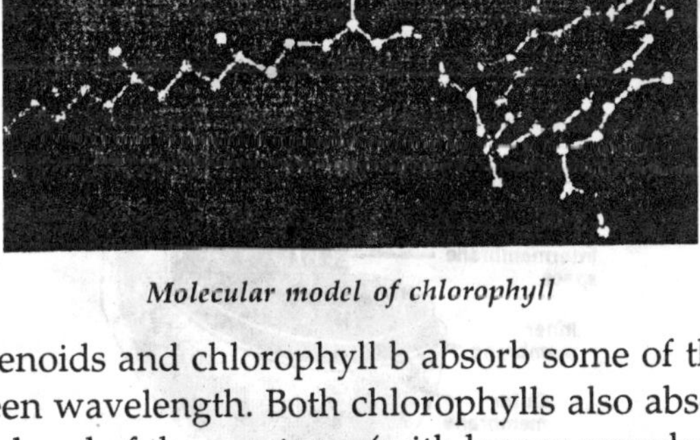

Molecular model of chlorophyll

Carotenoids and chlorophyll b absorb some of the energy in the green wavelength. Both chlorophylls also absorb in the orange-red end of the spectrum (with longer wavelengths and lower energy). The origins of photosynthetic organisms in the sea may account for this. Shorter wavelengths (with more energy) do not penetrate much below 5 metres deep in sea water. The ability to absorb some energy from the longer (hence more penetrating) wavelengths might have been an advantage to early photosynthetic algae that were not able to be in the upper (photic) zone of the sea all the time.

The action spectrum of photosynthesis is the relative effectiveness of different wavelengths of light at generating electrons. If a pigment absorbs light energy, one of three things will occur. Energy is dissipated as heat. The energy may be emitted immediately as a longer wavelength, a phenomenon known as fluorescence. Energy may trigger a chemical reaction, as in photosynthesis. Chlorophyll only triggers a chemical reaction when it is associated with proteins embedded in a membrane (as in a chloroplast) or the membrane infoldings found in photosynthetic prokaryotes such as cyanobacteria and prochlorobacteria.

Chloroplast and Photosynthetic Membranes: The thylakoid is the structural unit of photosynthesis. Both photosynthetic prokaryotes and eukaryotes have these flattened sacs/vesicles containing photosynthetic chemicals. Only eukaryotes have chloroplasts with a surrounding membrane. Thylakoids are stacked like pancakes in stacks known collectively as grana. The areas between grana are referred to as stroma. While the mitochondrion has two membrane systems, the chloroplast has three, forming three compartments.

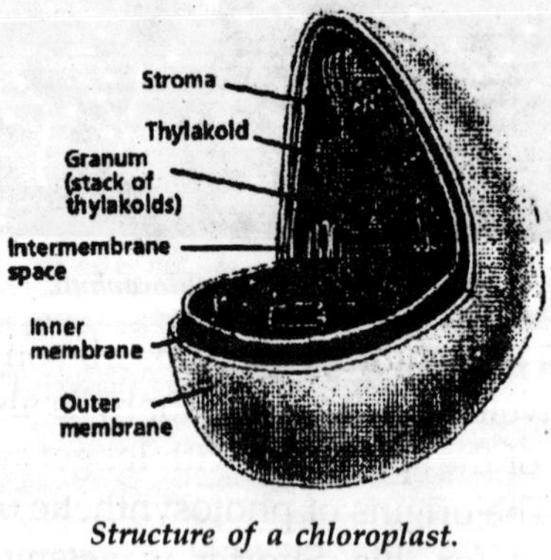

Structure of a chloroplast.

Stages of Photosynthesis

Photosynthesis is a two stage process. The first process is the Light Dependent Process (Light Reactions), requires the direct energy of light to make energy carrier molecules that are used in the second process. The Light Independent Process (or Dark Reactions) occurs when the products of the Light Reaction are used to form C-C covalent bonds of carbohydrates. The Dark Reactions can usually occur in the dark, if the energy carriers from the light process are present. Recent evidence suggests that a major enzyme of the Dark Reaction is indirectly stimulated by light, thus the term Dark Reaction is somewhat of a misnomer. The Light Reactions occur in the grana and the Dark Reactions take place in the stroma of the chloroplasts.

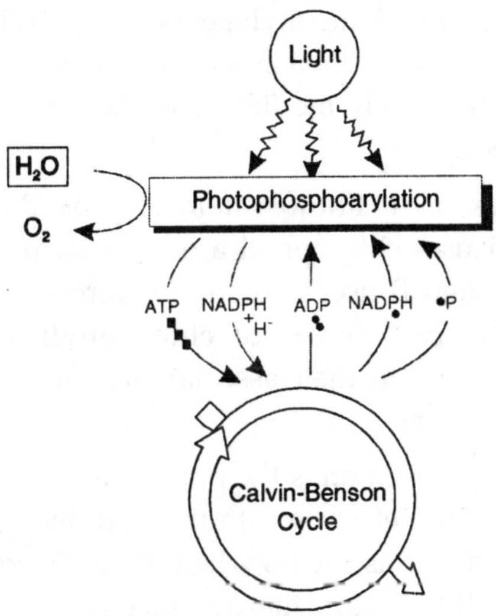

Two steps in the photosynthesis process.

Light Reactions: In the Light Dependent Processes (Light Reactions) light strikes chlorophyll a in such a way as to excite electrons to a higher energy state. In a series of reactions the energy is converted (along an electron transport process) into ATP and NADPH. Water is split in the process, releasing oxygen as a by-product of the reaction. The ATP and NADPH are used to make C-C bonds in the Light Independent Process (Dark Reactions). In the Light Independent Process, carbon dioxide from the atmosphere (or water for aquatic/marine organisms) is captured and modified by the addition of Hydrogen to form carbohydrates (general formula of carbohydrates is $[CH_2O]_n$).

The incorporation of carbon dioxide into organic compounds is known as carbon fixation. The energy for this comes from the first phase of the photosynthetic process. Living systems cannot directly utilise light energy, but can, through a complicated series of reactions, convert it into C-C bond energy that can be released by glycolysis and other metabolic processes. Photosystems are arrangements of chlorophyll and other pigments packed into thylakoids. Many Prokaryotes have only one photosystem, Photosystem II (so numbered because, while it was most likely the first to evolve, it was the second one discovered).

Eukaryotes have Photosystem II plus Photosystem I. Photosystem I uses chlorophyll a, in the form referred to as P700. Photosystem II uses a form of chlorophyll a known as P680. Both "active" forms of chlorophyll a function in photosynthesis due to their association with proteins in the thylakoid membrane.

Photophosphorylation is the process of converting energy from a light-excited electron into the pyrophosphate bond of an ADP molecule. This occurs when the electrons from water are excited by the light in the presence of P680.

The energy transfer is similar to the chemiosmotic electron transport occurring in the mitochondria. Light energy causes the removal of an electron from a molecule of P680 that is part of Photosystem II. The P680 requires an electron, which is taken from a water molecule, breaking the water into H^+ ions and O^{-2} ions. These O^{-2} ions combine to form the diatomic O_2 that is released.

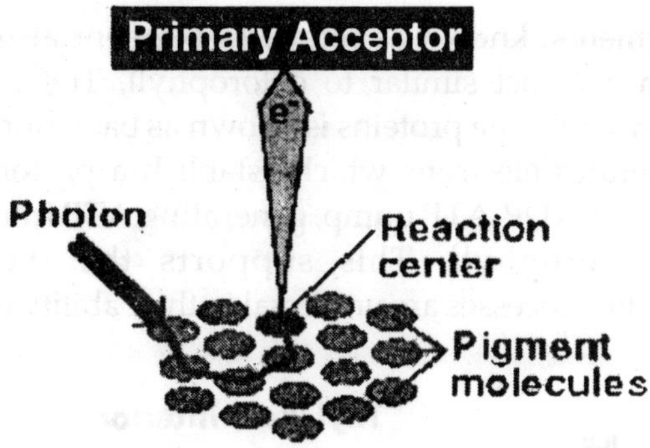

Action of a photosystem.

The electron is "boosted" to a higher energy state and attached to a primary electron acceptor, which begins a series of redox reactions, passing the electron through a series of electron carriers, eventually attaching it to a molecule in Photosystem I. Light acts on a molecule of P700 in Photosystem I, causing an electron to be "boosted" to a still higher potential. The electron is attached to a different primary electron acceptor (that is a different molecule from the one associated with Photosystem II). The electron is passed again through a series of redox reactions, eventually being attached to NADP+ and H+ to form NADPH, an energy carrier needed in the Light Independent Reaction.

The electron from Photosystem II replaces the excited electron in the P700 molecule. There is thus a continuous flow of electrons from water to NADPH. This energy is used in Carbon Fixation. Cyclic Electron Flow occurs in some eukaryotes and primitive photosynthetic bacteria. No NADPH is produced, only ATP. This occurs when cells may require additional ATP, or when there is no NADP+ to reduce to NADPH. In Photosystem II, the pumping to H ions into the thylakoid and the conversion of ADP + P into ATP is driven by electron gradients established in the thylakoid membrane.

Halobacteria, which grow in extremely salty water, are facultative aerobes, they can grow when oxygen is absent.

Purple pigments, known as retinal (a pigment also found in the human eye) act similar to chlorophyll. The complex of retinal and membrane proteins is known as bacteriorhodopsin, which generates electrons which establish a proton gradient that powers an ADP-ATP pump, generating ATP from sunlight without chlorophyll. This supports the theory that chemiosmotic processes are universal in their ability to generate ATP.

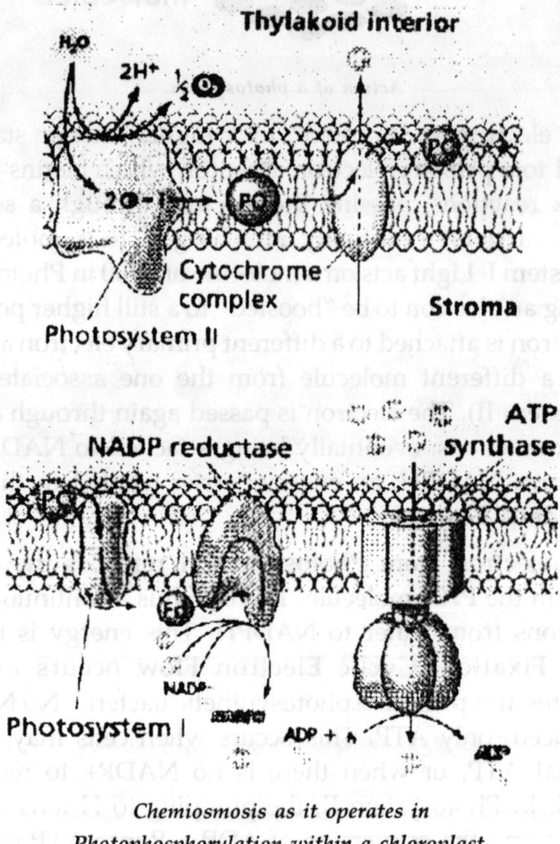

Chemiosmosis as it operates in
Photophosphorylation within a chloroplast.

Dark Reaction: Carbon-Fixing Reactions are also known as the Dark Reactions (or Light Independent Reactions). Carbon dioxide enters single-celled and aquatic autotrophs through no specialised structures, diffusing into the cells. Land plants must guard against drying out (desiccation) and so have evolved specialised structures known as stomata to allow gas to enter and leave the leaf. The Calvin Cycle occurs in the stroma of chloroplasts. Carbon dioxide is captured by the chemical ribulose biphosphate (RuBP). RuBP is a 5-C chemical. Six molecules of carbon dioxide enter the Calvin Cycle, eventually producing one molecule of glucose. The reactions in this process were worked out by Melvin Calvin.

The first stable product of the Calvin Cycle is phosphoglycerate (PGA), a 3-C chemical. The energy from ATP and NADPH energy carriers generated by the photosystems is used to attach phosphates to (phosphorylate) the PGA. Eventually there are 12 molecules of glyceraldehyde phosphate (also known as phosphoglyceraldehyde or PGAL, a 3-C), two of which are removed from the cycle to make a glucose. The remaining

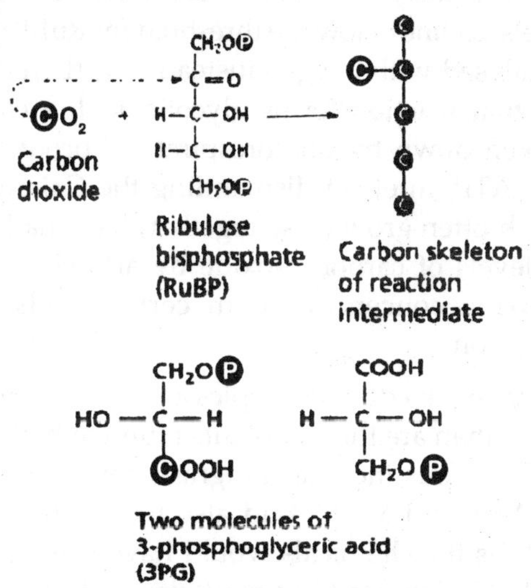

The first steps in the Calvin cycle.

PGAL molecules are converted by ATP energy to reform 6 RuBP molecules, and thus start the cycle again. Remember the complexity of life, each reaction in this process, as in Kreb's Cycle, is catalysed by a different reaction-specific enzyme.

C-4 Pathway: Some plants have developed a preliminary step to the Calvin Cycle (which is also referred to as a C-3 pathway), this preamble step is known as C-4. While most C-fixation begins with RuBP, C-4 begins with a new molecule, phosphoenolpyruvate (PEP), a 3-C chemical that is converted into oxaloacetic acid (OAA, a 4-C chemical) when carbon dioxide is combined with PEP. The OAA is converted to Malic Acid and then transported from the mesophyll cell into the bundle-sheath cell, where OAA is broken down into PEP plus carbon dioxide. The carbon dioxide then enters the Calvin Cycle, with PEP returning to the mesophyll cell. The resulting sugars are now adjacent to the leaf veins and can readily be transported throughout the plant.

The capture of carbon dioxide by PEP is mediated by the enzyme PEP carboxylase, which has a stronger affinity for carbon dioxide than does RuBP carboxylase When carbon dioxide levels decline below the threshold for RuBP carboxylase, RuBP is catalysed with oxygen instead of carbon dioxide. The product of that reaction forms glycolic acid, a chemical that can be broken down by photorespiration, producing neither NADH nor ATP, in effect dismantling the Calvin Cycle. C-4 plants, which often grow close together, have had to adjust to decreased levels of carbon dioxide by artificially raising the carbon dioxide concentration in certain cells to prevent photorespiration.

C-4 plants evolved in the tropics and are adapted to higher temperatures than are the C-3 plants found at higher latitudes. Common C-4 plants include crabgrass, corn, and sugar cane. Note that OAA and Malic Acid also have functions in other processes, thus the chemicals would have been present in all plants, leading scientists to hypothesise that C-4 mechanisms evolved several times independently in response to a similar environmental condition, a type of evolution known as convergent evolution.

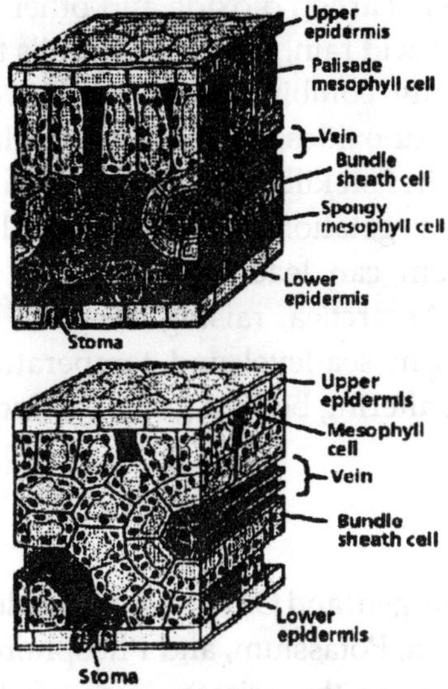

Leaf anatomy of a C3 (top) and C4 (bottom) plant.

Carbon Cycle: Plants may be viewed as carbon sinks, removing carbon dioxide from the atmosphere and oceans by fixing it into organic chemicals. Plants also produce some carbon dioxide by their respiration, but this is quickly used by photosynthesis. Plants also convert energy from light into chemical energy of C-C covalent bonds. Animals are carbon dioxide producers that derive their energy from carbohydrates and other chemicals produced by plants by the process of photosynthesis. The balance between the plant carbon dioxide removal and animal carbon dioxide generation is equalised also by the formation of carbonates in the oceans. This removes excess carbon dioxide from the air and water (both of which are in equilibrium with regard to carbon dioxide).

Fossil fuels, such as petroleum and coal, as well as more recent fuels such as peat and wood generate carbon dioxide when burned. Fossil fuels are formed ultimately by organic processes, and represent also a tremendous carbon sink. Human activity has greatly increased the concentration of carbon dioxide in air. This increase has led to global warming, an increase in temperatures around the world, the Greenhouse Effect.

The increase in carbon dioxide and other pollutants in the air has also led to acid rain, where water falls through polluted air and chemically combines with carbon dioxide, nitrous oxides, and sulphur oxides, producing rainfall with pH as low as 4. This results in fish kills and changes in soil pH which can alter the natural vegetation and uses of the land. The Global Warming problem can lead to melting of the ice caps in Greenland and Antarctica, raising sea-level as much as 120 metres. Changes in sea-level and temperature would affect climate changes, altering belts of grain production and rainfall patterns.

Balanced Diet

Carbon, Hydrogen, and Oxygen are considered the essential elements. Nitrogen, Potassium, and Phosphorous are obtained from the soil and are the primary macronutrients. Calcium, Magnesium, and Sulphur are the secondary macronutrients needed in lesser quantity. The micronutrients, needed in very small quantities and toxic in large quantities, include Iron, Manganese, Copper, Zinc, Boron, and Chlorine. A complete fertilizer provides all three primary macronutrients and some of the secondary and micronutrients. The label of the fertilizer will list numbers, for example 5-10-5, which refer to the per cent by weight of the primary macronutrients.

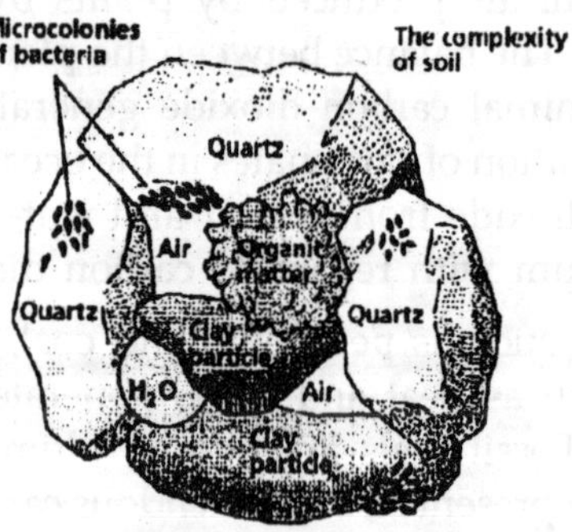

Structure of soil, indicating presence of bacteria, inorganic, and organic matter, water, and air.

Soil is weathered, decomposed rock and mineral (geological) fragments mixed with air and water. Fertile soil contains the nutrients in a readily available form that plants require for growth. The roots of the plant act as miners moving through the soil and bringing needed minerals into the plant roots.

Plants use these minerals in:

1. Structural components in carbohydrates and proteins
2. Organic molecules used in metabolism, such as the Magnesium in chlorophyll and the Phosphorous found in ATP
3. Enzyme activators like potassium, which activates possibly fifty enzymes
4. Maintaining osmotic balance

Plants need nitrogen for many important biological molecules including nucleotides and proteins. However, the nitrogen in the atmosphere is not in a form that plants can utilise. Many plants have a symbiotic relationship with bacteria growing in their roots: organic nitrogen as rent for space to live. These plants tend to have root nodules in which the nitrogen-fixing bacteria live. All the nitrogen in living systems was at one time processed by these bacteria, who took atmospheric nitrogen (N_2) and modified it to a form that living things could utilise.

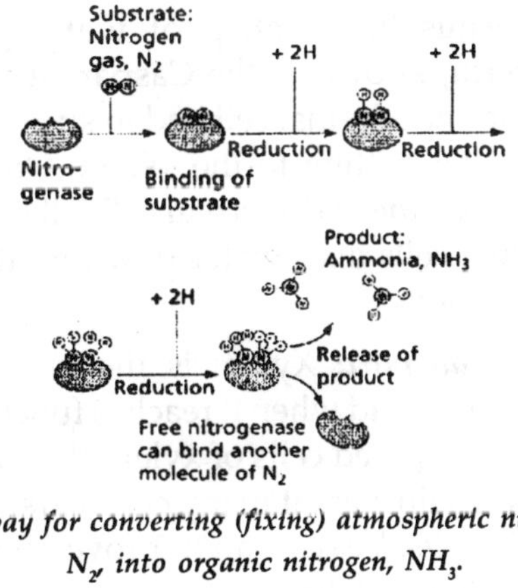

Pathway for converting (fixing) atmospheric nitrogen, N_2, into organic nitrogen, NH_3.

Not all bacteria utilise the above route of nitrogen fixation. Many that live free in the soil, utilise other chemical pathways.

Roots have extensions of the root epidemal cells known as root hairs. While root hairs greatly enhance the surface area (hence absorbtion surface), the addition of symbiotic mycorrhizae fungi vastly increases the area of the root for absorbing water and minerals from the soil.

Uptake of Water and Mineral: Animals have a circulatory system that transports fluids, chemicals, and nutrients around within the animal body. Some plants have an analogous system: the vascular system in vascular plants; trumpet hyphae in bryophytes. Root hairs are thin-walled extensions of the epidermal cells in roots. They provide increased surface area and thus more efficient absorption of water and minerals. Water and dissolved mineral nutrients enter the plant via two routes.

Water and selected solutes pass through only the cell membrane of the epidermis of the root hair and then through plasmodesmata on every cell until they reach the xylem: intracellular route (apoplastic). Water and solutes enter the cell wall of the root hair and pass between the wall and plasma membrane until they encounter the endodermis, a layer of cells that they must pass through to enter the xylem: extracellular route (symplastic).

The endodermis has a strip of water-proof material (containing Suberin) known as the Casparian strip that forces water through the endodermal cell and in such a way regulates the amount of water getting to the xylem. Only when water concentrations inside the endodermal cell fall below that of the cortex parenchyma cells does water flow into the endodermis and on into the xylem.

Xylem and Transport: Xylem is the water transporting tissue in plants that is dead when it reaches functional maturity. Tracheids are long, tapered cells of xylem that have end plates on the cells that contain a great many crossbars. Tracheid walls are festooned with pits. Vessels, an improved form of tracheid, have no (or very few) obstructions (crossbars) on the top or bottom of the cell. The functional diameter of vessels is greater than that of Tracheids. Water is pulled up the xylem by the force of transpiration, water loss from leaves. Mature corn plants can each transpire four gallons of water per week. Transpiration rates in arid-region plants can be even higher.

Water molecules are hydrogen bonded to each other. Water lost from the leaves causes diffusion of additional water molecules out of the leaf vein xylem, creating a tug on water molecules along the water columns within the xylem. This "tug" causes water molecules to rise up from the roots to eventually the leaves. The loss of water from the root xylem allows additional water to pass through the endodermis into the root xylem.

Cohesion is the ability of molecules of the same kind to stick together. Water molecules are polar, having slight positive and negative sides, which causes their cohesion. Inside the xylem, water molecules are in a long chain extending from the roots to the leaves. Adhesion is the tendency of molecules of different kinds to stick together. Water sticks to the cellulose molecules in the walls of the xylem, counteracting the force of gravity and aiding the rise of water within the xylem.

Cohesion-Adhesion Theory: Transpiration exerts a pull on the water column within the xylem. The lost water molecules are replaced by water from the xylem of the leaf veins, causing a tug on water in the xylem. Adhesion of water to the cell walls of the xylem facilitates movement of water upward within the xylem. This combination of cohesive and adhesive forces is referred to as the Cohesion-Adhesion Theory.

Guard Cells: In most environments, the water concentration outside the leaf is less than that inside the leaf, causing a loss of water through openings in the leaf known as stomata (singular = stoma). Guard cells are crescent-shaped cells of the epidermis that flank the stoma and regulate the size of the opening. Together, the guard cells and stoma comprise the stomatal apparatus.

The inner wall of the guard cell is thicker than the rest of the wall. When a guard cell takes up potassium ions, water moves into the cell, causing the cell to become turgid and swell, opening the stoma. When the potassium leaves the guard cell, the water also leaves, causing plasmolysis of the cells, and a closing of the stoma. Stomata occupy 1 per cent of the leaf surface, but account for 90 per cent of the water lost in transpiration.

Nutrient Storage

Plants make sugar by photosynthesis, usually in their leaves. Some of this sugar is directly used for the metabolism of the plant, some for the synthesis of proteins and lipids, some stored as starch. Other parts of the plant also need energy but are not photosynthetic, such as the roots. Food must therefore be transported in from a source, an action accomplished by the phloem tissue.

Phloem, Sugar, and Translocation: Phloem consists of several types of cells: sieve tube cells (aka sieve elements), companion cells, and the vascular parenchyma. Sieve cells are tubular cells with endwalls known as sieve plates. Most lose their nuclei but remain alive, leaving an empty cell with a functioning plasma membrane. Companion cells load sugar into the sieve element (sieve elements are connected into sieve tubes). Fluids can move up or down within the phloem, and are translocated from one place to another. Sources are places where sugars are being produced. Sinks are places where sugar is being consumed or stored.

Food moves through the phloem by a Pressure-Flow Mechanism. Sugar moves (by an energy-requiring step) from a source (usually leaves) to a sink (usually roots) by osmotic pressure. Translocation of sugar into a sieve element causes water to enter that cell, increasing the pressure of the sugar/ water mix (phloem sap). The pressure causes the sap to flow towards an area of lower pressure, the sink. In the sink, the sugar is removed from the phloem by another energy-requiring step and usually converted into starch or metabolised.

6

Transport System for Water (Plants)

Plants have a transport system, in some ways similar to an animal's blood circulatory system. However, it is rather different in several important ways. For example, there is no pump like the heart, no circulating cells and liquids do not continuously move round and round. The substances which are transported - mineral salts (ions) from the soil, and the products of photosynthesis from the leaves - are dissolved in water (as an "aqueous solution"). The transport system basically consists of 2 types of conducting tissue, each of which is made from cells which have been modified for their special purpose. Some cells die as a result of this modification, and they may also lose some of their internal components.

Water and mineral salts enter a plant through special cells called root hair cells. The water is taken up by a special form of diffusion called osmosis, but the mineral salts (ions) may also be taken up by active transport which uses some of the plant's energy to concentrate them. Xylem vessels carry water and minerals on into the plant, but only in an upward direction. These vessels form a continuous "pipe" from the root to the stem and leaves.

Xylem consists of fairly wide tubes with thick walls. These xylem vessel elements were originally cells which died due to the buildup of a tough substance called lignin in their cell walls, and then lost their end walls and cytoplasm. Phloem tubes carry sugars (sucrose is more likely than glucose) and other "foods" such as amino acids produced as a result of photosynthesis. The direction of this movement can be either upwards or downwards from the leaves.

Phloem consists of cells called sieve tubes because their end walls have several small holes with strands of cytoplasm running vertically through them. Sieve tube cells have no nuclei, but they are not dead, and their activity is controlled by companion cells alongside. These 2 conducting tissues are to be found in roots, stems and also in leaves. However, the pattern of their arrangement differs in each.

Plant-water Relationship

Water is essential in the plant environment for a number of reasons. Water transports minerals through the soil to the roots where they are absorbed by the plant. Water is also the principal medium for the chemical and biochemical processes that support plant metabolism. Under pressure within plant cells, water provides physical support for plants. It also acts as a solvent for dissolved sugars and minerals transported throughout the plant. In addition, evaporation within intercellular spaces provides the cooling mechanism that allows plants to maintain the favourable temperatures necessary for metabolic processes.

Water is transported throughout plants almost continuously. There is a constant movement of water from the soil to the roots, from the roots into the various parts of the plant, then into the leaves where it is released into the atmosphere as water vapour through the stomata (small openings in the leaf surfaces). This process is called transpiration. Combined with evaporation from the soil and wet plant surfaces the total water loss to the atmosphere is called evapotranspiration. Guard cells which are found on both sides of the stoma control its opening and closing.

Stomata can be found on one (typically underside) or both sides of a leaf depending on plant species.

The major economic consequence of insufficient water in agricultural crops is yield reduction. When too little water is available in the root zone, the plant will reduce the amount of water lost through transpiration by partial or total stomatal closure. This results in decreased photosynthesis since the CO_2 required for this process enters the plant through the stomata. Decreased photosynthesis reduces biomass production and results in decreased yields.

Transport of Water

The ancestors of modern day plants evolved in water and it is no surprise that the internal environment of a plant is 80-90 per cent water. The relative humidity inside plants is always higher than 95 per cent and usually higher than 98 per cent. As land plants must remain hydrated to this level if they are to continue growing, they have evolved a number of mechanisms, such as waxy cuticles, to ensure that they do not dry out. The largest water problem a land plant faces is trying to grow, because this involves the fundamental process on earth, photosynthesis.

The sugar produced in photosynthesis is the building block for all plant growth and therefore all higher forms of life on earth. The major water problem for plants is not the water used in photosynthesis, and subsequent growth (less than 2 per cent of total plant water use), but the water that evaporates out of the leaf as it captures CO_2 from an atmosphere that only contains 0.035 per cent CO_2. The simultaneous movement of

CO_2 in and out of a leaf occurs by a process known as diffusion, which is the independent movement of individual molecules. This gas exchange in the leaf does not occur all over the leaf, but rather through small adjustable parts called stomata, which usually make up only 1 per cent of the leaf surface area.

The evaporation of water out of the plant to the atmosphere is called transpiration. The relative amounts of photosynthesis (CO_2 in) and transpiration (H_2O out) occurring at any one time depends on amounts (vapour pressure) of CO_2 and H_2O in the atmosphere inside and outside the leaf respectively.

The plant can control the loss of water from its leaves by varying the aperture of its stomata (like a tap). However, if a plant restricts the flow of water vapour out of its leaf it automatically restricts the flow of CO_2 into the leaf for photosynthesis. This is known as the transpiration compromise or the plant version of 'there are no free lunches'.

Transpiration

Water flows through plants in well constructed interconnected system of pipes called xylem, vessel elements and tracheids at speeds of one to several metres per hour. Rates are usually faster in woody plants than herbaceous plants, such as grasses, because the xylem vessels are bigger. The important feature of this plumbing system is that it is closed at both ends, the roots and the leaves, by cell walls, cellulose structures like ultra thin tissue paper. The roots are the water absorbing organs in the soil. The ability of a plant to absorb water from the soil depends on the number, rather than the size of the roots and where they are distributed in the soil.

Even root distribution is important for both water and nutrient uptake, because of the relative rates of movement of water and nutrients in the soil and in the plant xylem. Root hairs are more important for nutrient uptake than water because they do not contain xylem. Although water moves in saturated permeable soils at rates of between 0.01 to 0.2 metres per hour, the rate of movement in unsaturated soils (which is most of the time) is 10 to 100 times slower. Remembering that water moves in the xylem of grasses at about 1 metre per hour, we can appreciate that water moves 50 to 1000 times SLOWER in soil than roots.

Useful quantities of water will not move more than a few centimetres in unsaturated soils, therefore if you wish the plant to withdraw H_2O from a soil layer it must have roots in it. The xylem becomes of necessarily larger diameter, as it collects water from smaller roots to larger roots, than the stem. The reverse occurs in the leaves where successively smaller xylem elements distribute water to the leaves, then the veins within the leaves, finally ending in small groups of cells surrounding the stomata.

The end of each small xylem element is closed by a cell wall which is mainly cellulose and acts like an ultra think tissue paper. The water then seeps along the cell walls to the sub-stomata cavity, where it evaporates and moves through the stomata to the outside atmosphere. The important point about the xylem pathway is that is a continuous column of water, not broken by bubbles, and contained in a cellulose pipe whose walls are porous in places.

The rate of water movement through the plant depends on the rate of evaporation from the leaf, it is strongly influenced by environmental conditions:

1. Air temperature
2. Relative humidity
3. Wind speed
4. Radiation intensity

The second mechanism that contributes to water movement is an osmotic pumping mechanism that operates in the roots. In this mechanism, nutrients organic and inorganic, in the cells of the root attract water to move into the root, in the same way that salt in your salt shaker attracts water. This causes pressure to build up in the root, forcing water up the xylem. Although this osmotic pumping mechanism operates continually, it does not have the capacity to supply the flow of water needed for transpiration during daylight hours. However, it is the major mechanism operating at night and the one responsible for developing the root pressure that results in the small droplets of water we often see on the edges of leaves in the morning. These two mechanisms of water transport work in plants, with the evaporative mechanism being dominant in the transpiration process and the osmotic pump taking over at night.

The plant can exert strong control over the rate of water loss through varying stomatal apertures. Stomates respond to environmental factors such as light, temperature and relative humidity to ensure that the plant is making the best use of its resources to photosynthesise and grow. This usually means that under well watered conditions stomates respond to light, being open during the day and closed at night.

When plant roots can no longer absorb enough water to satisfy transpiration from its leaves, it begins to dehydrate. At this point the stomates can over-ride photosynthesis and close down for certain parts of the day to avoid further water loss. Under extensive water stress where the plant is unable to rehydrate at night, the stomates will remain closed all day. While this is very effective in preventing water loss, it prevents the absorption of CO_2 and therefore photosynthesis. The plant cannot live indefinitely on its internal reserves.

When transpiration exceeds water absorption by the roots, the plant dehydrates. This usually happens each day with the plant rehydrating again each night. As the soil dries out this rehydration is not complete, resulting in the plant becoming water stressed.

Water stress affects different plant processes such as growth, photosynthesis and transpiration in different ways. The growth of leaves and stems is much more sensitive to water stress than photosynthesis and transpiration.

Therefore it is possible to restrict the growth of a plant by water stressing it, without underlying affecting its photosynthesis and survival. Through an understanding of the water availability in the soil and the atmospheric demand, water can be used as a potent management variable to manipulate the growth and maintenance of turf grasses.

Membranes

Each cell is encircled by a membrane and most cells are equipped with an extensive intracellular membrane system. Membranes fence the cell's interior off from its surrounding. They let water, certain ions and substrates into the cell, are vital for the cell's energy budget and they excrete substances. They protect the cell. Without them would the cell contents diffuse into their surroundings, information containing molecules would be lost and the metabolic pathways would seek a thermal equilibrium: the death of each living system.

Intracellular membranes separate compartments from each other. Some surround vacuoles, others mitochondria, chloroplasts and the nucleus. Compartments are reaction areas clearly separated spatially. All biological membranes have a great deal in common as regards organisation and composition:

1. They have an average thickness of 70 Å.
2. They consist mainly of lipids and proteins.
3. They act as a permeability barriers and are permeable only for selected substances.
4. When fragmented do they always reform as vesicles.

Phospholipids are the most common molecules in membranes. They consist of two parts, a hydrophilic (polar) head group (the phosphate group with the attached rest R) and a hydrophobic (lipophilic) fatty acid chain. This structure lends a tendency to form organised structures and monomolecular layers whose degree of organisation is dependent on the length of the fatty acid residues and the number of double bonds to phospholipids. The hydrophobic parts of the molecule would point towards the inside of the double layer, the hydrophilic towards its outside.

Diffusion

Atoms, ions and molecules are constantly moving, colliding and thus changing their direction. In an ideal gas move the single particles independently of one another. The particle distribution is not influenced by these movements and no predominant or preferred direction exists. The velocity depends on the temperature and increases when the temperature rises. It is therefore also spoken of thermal movements (T). The particle movements within liquids are more restricted but the movements within a solution are subject to the same criteria as those in an ideal gas. Often can the plasma of a cell, too, be regarded as such a solution.

If a concentrated solution of any substance like, for example, sugar, is layered with water do the layers at first stay clearly separated. Since we assumed that dissolved particles move in a strictly statistical manner, will at first single particles cross from one to the other liquid and their number will increase with time [f (t)]. Some of them, though not all will return and a net flow of particles moving in the direction formerly free of sugar molecules will develop.

Particle movements from high to lower concentrations are called diffusion. Only after an even distribution is achieved can no further net movements be observed, the system has reached an equilibrium. The number of particles that cover a certain distance by diffusion is dependent on the cross-section's area of the chosen vessel (F) and a substance constant (D).

Diffusion is very quick over short distances but extremely slow at long ones. It is the square of the distance that has an influence on the formula and it is proportional to the available area. Diffusion is important when regarding the molecular movements within cells or between neighbouring cells. But it losses its significance when regarding whole tissues and ceases to have any effect at all when, for example, thinking of the assimilate transport from leaves to roots.

Permeability

Permeability is the term that describes the diffusion of particles through membranes. It is unimportant whether we regard natural or artificial membranes (like plastic film). We will in a rough approximation assume that the membrane equals a solvent layer of the thickness in order to obtain a quantitative statement.

Biological membranes are not equally permeable for all substances. They are selectively permeable, i.e. the membranes can be permeated by a substance A but not by a substance B. Strictly speaking are they only permeable for some gases like oxygen, nitrogen, carbon dioxide, etc., because only these can permeate a membrane by free diffusion. Diffusion is strictly subject to the laws of thermodynamics. Molecules are characterised by specific properties like size, ability to be converted into a ion or solubility within a membrane. Such lipophilic (fat soluble) molecules can cross the membrane more easily than hydrophilic ones.

By a specific arrangement of membrane components pores or channels can be formed that can be selectively permeated by certain ions and molecules. Their permeability is on one hand dependent on the diameter and on the other hand on the electric charge of the pore. Small anions like chloride can easily pass positively charged pores. Cations are detained. Such a state is also called a restricted diffusion.

Numerous small molecules like sugar and amino acids seem to be preferentially transported through membranes. But their permeation kinetics does not follow the formula discussed; it resembles more an enzyme-substrate turnover kinetic. Such a kinetic is known as facilitated diffusion. The permeability is selective and dependent on the presence of specific mem brane-bound transport molecules with an affinity to a selected group of chemically related substances. The substrates do not pass the membrane by themselves but are transported as a substrate-carrier complex

Osmosis

Biological membranes are usually permeable for water and gases like oxygen, nitrogen, carbon dioxide and others and semi-permeable for some ions, sugar and other small molecules. They are, with some exceptions, impermeable for large molecules. When disregarding mediated transport (transport occurring through the action of specific carrier molecules) is the permeability of the membrane dependent on its pore size. Osmosis is the net flow of water through a semi-permeable membrane that is driven by a difference in concentration of solute on either side.

A solution that is separated from another solution by a semi-permeable membrane can have three osmotic states:

— In an isotonic solution is the pressure at both sides of the membrane the same.

— A hypotonic solution has a lesser number of solute particles than the solution to which it is compared, while

— A hypertonic solution has a higher number of solute particles. At equilibrium a solution always isotonic.

Each cell has constantly to cope with osmotic phenomenons. Cells without walls that live in aqueous solutions are usually hypertonic. They are subject to a continuous influx of water that results in a pressure on the inside of the membrane. Some ciliates (like Paramecium) and flagellates (Euglena) have a special organ, a pulsating vacuole, that pumps the water out again and has to be supplied with energy. Red blood cells (erythrocytes) occur normally in isotonic surroundings (blood plasma). The cells burst if the blood is diluted with water since their membranes cannot face the osmotic pressure within the cells. Blood is therefore always diluted with an isotonic (0.9 per cent) NaCl-solution.

The situation in plant cells is different in that they are enclosed by a rigid cell wall. In a hypotonic solution can they only take up water until the inner and the outer water potential are equal. The solution at the inside of the plant cell is subject to an additional hydrostatic pressure. Plant cells store ions, sugars, organic and amino acids and other substances in considerable concentrations in their vacuoles. The solutes cause an influx of water. In this way can plant cells build up a large positive internal pressure, the turgor pressure. It has a decisive influence on the maintenance of the rigidity and stability of plant tissues. Each cell exerts a pressure on its neighbouring cells. The pressures add up to a large tissue tension. Plants that loose water wilt because their turgor pressure decreases. The stability of the tissues cannot be maintained. As long as the cells are still living the turgor pressure can be fixed again, a phenomenon common in everyday life: wilted plants recover after watering.

Active Transport

Participation of a carrier molecule in the active transport of certain molecules through the membrane. The transport process is specific and energy-consuming. The conformation of the carrier changes during transport (ping-pong mechanism). Active transport is the movement of a molecule across a membrane or another barrier that is driven by energy other than stored in the concentration gradient or the electrochemical gradient of the transported molecule. This type of transport requires usually the expenditure of ATP and the help of specific transport proteins. In this way even large molecules can be channelled through the membrane.

Active transport can only occur at intact, closed membranes. Such membranes can envelop very different compartments, like the whole cell, vesicles, the vacuole, the mitochondrial matrix, the inner thylacoid space of the chloroplasts, etc. As a result of active transport can ions and metabolites be concentrated within the respective compartment or the cell and the steady state of the metabolism can be kept constant despite of large fluctuations in the external medium's composition. Ions, especially potassium, calcium, magnesium and phosphate have an important part in the regulation of the metabolism. The transport direction is thermodynamically determined by coupling the transport with a gradient, usually an electron gradient. The direction can be reversed if the appropriate substrate concentrations are chosen.

As a consequence of the respiratory chain and of photosynthesis are protons channelled out of the mitochondrial matrix and the thylacoids of the chloroplasts, respectively. At the same time causes the efflux of protons a translocation of electric charges. That way is both a chemical proton gradient and an electric potential generated. The tendency of the protons (H+) to return to the inside of the compartment is therefore quite large. It is called the proton motive force.

ATP hydrolysis is actually the reversion of oxidative or Photophosphorylation. The ATP synthethase works the other way round: protons (and other cations like potassium) are actively transported out of the compartment (proton pump) and build as a consequence a pH gradient and a membrane potential up. These two forces again drive the translocation of other molecules (or ions) via antiport or symport.

The membrane-bound proteins needed for active transport are without exception oligomerous complexes, i.e. they consist of several protein subunits. They are often termed pumps, too. A reversible ATP synthethase (able to hydrolyse ATP) that gets its energy from the electrochemical potential difference between inside and outside is among the most important components of the whole complex. The transport kinetics of the pumps can -similar to those of facilitated diffusion - be compared with enzyme kinetics. The proton complexes work analogous to enzymes and are therefore often also classified as permeases. Proteins of animal membranes (like erythrocytes) and micro-organisms have been studied especially extensively.

The number of polypeptide chains that make up a complex, the molecular weight and the number of ions or molecules transported per time unit is often known. It is also known, that many pumps work only when in their specific surrounding, in other words when in close vicinity to the right phospholipids. And finally is it known that pumps are integrated into the membrane in always the same orientation because transport is always directed. But in no case is the exact tertiary or the quaternary structure of the complex known and only these structures would explain, why a specific molecule or ion and no other, similar ones are transported.

The best-known and best-studied ion pumps are the sodium-potassium-and the calcium pump. The sodium/ potassium pump was isolated from the most different types of membranes. It is generally known that cells contain more potassium than sodium ions. This is also true for plants living in saline (sodium containing) soils. Such plants are also said to be halophilic. Especially in the cells of halophilic plants was a very large activity of sodium / potassium exchanges detected. The cells invest a considerable amount of energy in order to maintain within the cytosol as low a concentration as possible. Part of the sodium is pumped into the vacuole where it does not interfere but helps to maintain the osmotic pressure.

A number of secondary plant products like the glycosides oubain (strophanthin) and digitonin inhibit the sodium/ potassium pump selectively. The calcium pump located especially in intracellular and organelle-membranes is of no lesser importance. The calcium concentrations within the different ccmpartments differ by scales.

Circulatory System in Animals

Living things must be capable of transporting nutrients, wastes and gases to and from cells. Single-celled organisms use their cell surface as a point of exchange with the outside environment. Multicellular organisms have developed transport and circulatory systems to deliver oxygen and food to cells and remove carbon dioxide and metabolic wastes. Sponges are the simplest animals, yet even they have a transport system. Seawater is the medium of transport and is propelled in and out of the sponge by ciliary action. Simple animals, such as the hydra and planaria, lack specialised organs such as hearts and blood vessels, instead using their skin as an exchange point for materials. This, however, limits the size an animal can attain. To become larger, they need specialised organs and organ systems.

Multicellular animals do not have most of their cells in contact with the external environment and so have developed circulatory systems to transport nutrients, oxygen, carbon dioxide and metabolic wastes. Components of the circulatory system include

- — blood: a connective tissue of liquid plasma and cells
- — heart: a muscular pump to move the blood
- — blood vessels: arteries, capillaries and veins that deliver blood to all tissues

There are several types of circulatory systems. The open circulatory system is common to molluscs and arthropods. Open circulatory systems (evolved in insects, mollusks and other invertebrates) pump blood into a hemocoel with the blood diffusing back to the circulatory system between cells. Blood is pumped by a heart into the body cavities, where tissues are surrounded by the blood. The resulting blood flow is sluggish.

Vertebrates, and a few invertebrates, have a closed circulatory system. Closed circulatory systems (evolved in echinoderms and vertebrates) have the blood closed at all times within vessels of different size and wall thickness. In this type of system, blood is pumped by a heart through vessels, and does not normally fill body cavities. Blood flow is not sluggish. Hemoglobin causes vertebrate blood to turn red in the presence of oxygen; but more importantly hemoglobin molecules in blood cells transport oxygen. The human closed circulatory system is sometimes called the cardiovascular system.

Vertebrate Cardiovascular System: The vertebrate cardiovascular system includes a heart, which is a muscular pump that contracts to propel blood out to the body through arteries, and a series of blood vessels. The upper chamber of the heart, the atrium (pl. atria), is where the blood enters the heart. Passing through a valve, blood enters the lower chamber, the ventricle. Contraction of the ventricle forces blood from the heart through an artery. The heart muscle is composed of cardiac muscle cells.

Arteries are blood vessels that carry blood away from heart. Arterial walls are able to expand and contract. Arteries have three layers of thick walls. Smooth muscle fibres contract, another layer of connective tissue is quite elastic, allowing the arteries to carry blood under high pressure. The aorta is the main artery leaving the heart. The pulmonary artery is the only artery that carries oxygen-poor blood. The pulmonary artery carries deoxygenated blood to the lungs. In the lungs, gas exchange occurs, carbon dioxide diffuses out, oxygen diffuses in. Arterioles are small arteries that connect larger arteries with capillaries. Small arterioles branch into collections of capillaries known as capillary beds.

Capillaries are thin-walled blood vessels in which gas exchange occurs. In the capillary, the wall is only one cell layer thick. Capillaries are concentrated into capillary beds. Some capillaries have small pores between the cells of the capillary wall, allowing materials to flow in and out of capillaries as well as the passage of white blood cells. Nutrients, wastes, and hormones are exchanged across the thin walls of capillaries. Capillaries are microscopic in size, although blushing is one manifestation of blood flow into capillaries. Control of blood flow into capillary beds is done by nerve-controlled sphincters.

The circulatory system functions in the delivery of oxygen, nutrient molecules, and hormones and the removal of carbon dioxide, ammonia and other metabolic wastes. Capillaries are the points of exchange between the blood and surrounding tissues. Materials cross in and out of the capillaries by passing through or between the cells that line the capillary.

The extensive network of capillaries in the human body is estimated at between 50,000 and 60,000 miles long. Thoroughfare channels allow blood to bypass a capillary bed. These channels can open and close by the action of muscles that control blood flow through the channels.

Blood leaving the capillary beds flows into a progressively larger series of venules that in turn join to form veins. Veins carry blood from capillaries to the heart. With the exception of the pulmonary veins, blood in veins is oxygen-poor. The pulmonary veins carry oxygenated blood from lungs back to the heart. Venules are smaller veins that gather blood from capillary beds into veins. Pressure in veins is low, so veins depend on nearby muscular contractions to move blood along. The veins have valves that prevent back-flow of blood.

Ventricular contraction propels blood into arteries under great pressure. Blood pressure is measured in mm of mercury; healthy young adults should have pressure of ventricular systole of 120mm, and 80 mm at ventricular diastole. Higher pressures (human 120/80 as compared to a 12/1 in lobsters) mean the volume of blood circulates faster (20 seconds in humans, 8 minutes in lobsters).

As blood gets farther from the heart, the pressure likewise decreases. Each contraction of the ventricles sends pressure through the arteries. Elasticity of lungs helps keep pulmonary pressures low. Systemic pressure is sensed by receptors in the arteries and atria. Nerve messages from these sensors communicate conditions to the medulla in the brain. Signals from the medulla regulate blood pressure.

Vertebrate Vascular Systems: Humans, birds, and mammals have a 4-chambered heart that completely separates oxygen-rich and oxygen-depleted blood. Fish have a 2-chambered heart in which a single-loop circulatory pattern takes blood from the heart to the gills and then to the body. Amphibians have a 3-chambered heart with two atria and one ventricle. A loop from the heart goes to the pulmonary capillary beds, where gas exchange occurs. Blood then is returned to the heart. Blood exiting the ventricle is diverted, some to the pulmonary circuit, some to systemic circuit. The disadvantage of the three-chambered heart is the mixing of oxygenated and deoxygenated blood. Some reptiles have partial separation of the ventricle. Other reptiles, plus, all birds and mammals, have a 4-chambered heart, with complete separation of both systemic and pulmonary circuits.

Heart: The heart is a muscular structure that contracts in a rhythmic pattern to pump blood. Hearts have a variety of forms: chambered hearts in mollusks and vertebrates, tubular hearts of arthropods, and aortic arches of annelids. Accessory hearts are used by insects to boost or supplement the main heart's actions. Fish, reptiles, and amphibians have lymph hearts that help pump lymph back into veins. The basic vertebrate heart, such as occurs in fish, has two chambers. An auricle is the chamber of the heart where blood is received from the body. A ventricle pumps the blood it gets through a valve from the auricle out to the gills through an artery.

Amphibians have a three-chambered heart: two atria emptying into a single common ventricle. Some species have a partial separation of the ventricle to reduce the mixing of oxygenated (coming back from the lungs) and deoxygenated blood (coming in from the body). Two sided or two chambered hearts permit pumping at higher pressures and the addition of the pulmonary loop permits blood to go to the lungs at lower pressure yet still go to the systemic loop at higher pressures.

Establishment of the four-chambered heart, along with the pulmonary and systemic circuits, completely separates oxygenated from deoxygenated blood. This allows higher the metabolic rates needed by warm-blooded birds and mammals.

The human heart is a two-sided, 4 chambered structure with muscular walls. An atrioventricular (AV) valve separates each auricle from ventricle. A semi-lunar (also known as arterial) valve separates each ventricle from its connecting artery. The heart beats or contracts 70 times per minute. The human heart will undergo over 3 billion contraction cycles during a normal lifetime. The cardiac cycle consists of two parts: systole (contraction of the heart muscle) and diastole (relaxation of the heart muscle). Atria contract while ventricles relax. The pulse is a wave of contraction transmitted along the arteries. Valves in the heart open and close during the cardiac cycle. Heart muscle contraction is due to the presence of nodal tissue in two regions of the heart. The SA node (sinoatrial node) initiates heartbeat. The AV node (atrioventricular node) causes ventricles to contract. The AV node is sometimes called the pacemaker since it keeps heartbeat regular. Heartbeat is also controlled by the autonomic nervous system.

Blood flows through the heart from veins to atria to ventricles out by arteries. Heart valves limit flow to a single direction. One heartbeat, or cardiac cycle, includes atrial contraction and relaxation, ventricular contraction and relaxation, and a short pause. Normal cardiac cycles (at rest) take 0.8 seconds. Blood from the body flows into the vena cava, which empties into the right atrium. At the same time, oxygenated blood from the lungs flows from the pulmonary vein into the left atrium.

The muscles of both atria contract, forcing blood downward through each AV valve into each ventricle. Diastole is the filling of the ventricles with blood. Ventricular systole opens the SL valves, forcing blood out of the ventricles through the pulmonary artery or aorta. The sound of the heart contracting and the valves opening and closing produces a characteristic "lub-dub" sound. Lub is associated with closure of the AV valves, dub is the closing of the SL valves. Human heartbeats originate from the sinoatrial node (SA node) near the right atrium.

Modified muscle cells contract, sending a signal to other muscle cells in the heart to contract. The signal spreads to the atrioventricular node (AV node). Signals carried from the AV node, slightly delayed, through bundle of His fibres and Purkinjie fibres cause the ventricles to contract simultaneously.

The Vascular System: Two main routes for circulation are the pulmonary (to and from the lungs) and the systemic (to and from the body). Pulmonary arteries carry blood from the heart to the lungs. In the lungs gas exchange occurs. Pulmonary veins carry blood from lungs to heart. The aorta is the main artery of systemic circuit. The vena cavae are the main veins of the systemic circuit. Coronary arteries deliver oxygenated blood, food, etc. to the heart.

Animals often have a portal system, which begins and ends in capillaries, such as between the digestive tract and the liver. Fish pump blood from the heart to their gills, where gas exchange occurs, and then on to the rest of the body. Mammals pump blood to the lungs for gas exchange, then back to the heart for pumping out to the systemic circulation. Blood flow is only one directional.

Blood: Plasma is the liquid component of the blood. Mammalian blood consists of a liquid (plasma) and a number of cellular and cell fragment components. Plasma is about 60 per cent of a volume of blood; cells and fragments are 40 per cent. Plasma has 90 per cent water and 10 per cent dissolved materials including proteins, glucose, ions, hormones, and gases. It acts as a buffer, maintaining pH near 7.4. Plasma contains nutrients, wastes, salts, proteins, etc. Proteins in the blood aid in transport of large molecules such as cholesterol.

Red blood cells, also known as erythrocytes, are flattened, doubly concave cells about 7 um in diameter that carry oxygen associated in the cell's hemoglobin. Mature erythrocytes lack a nucleus. They are small, 4 to 6 million cells per cubic millimetre of blood, and have 200 million hemoglobin molecules per cell. Humans have a total of 25 trillion (about 1/3 of all the cells in the body). Red blood cells are continuously manufactured in red marrow of long bones, ribs, skull, and vertebrae. Life-span of an erythrocyte is only 120 days, after which they are destroyed in liver and spleen. Iron from hemoglobin is recovered and reused by red marrow.

The liver degrades the heme units and secretes them as pigment in the bile, responsible for the colour of faces. Each second 2 million red blood cells are produced to replace those taken out of circulation. White blood cells, also known as leukocytes, are larger than erythrocytes, have a nucleus, and lack hemoglobin. They function in the cellular Immune response. White blood cells (leukocytes) are less than 1 per cent of the blood's volume. They are made from stem cells in bone marrow. There are five types of leukocytes, important components of the immune system. Neutrophils enter the tissue fluid by squeezing through capillary walls and phagocytosing foreign substances. Macrophages release white blood cell growth factors, causing a population increase for white blood cells. Lymphocytes fight infection.

T-cells attack cells containing viruses. B-cells produce antibodies. Antigen-antibody complexes are phagocytised by a macrophage. White blood cells can squeeze through pores in the capillaries and fight infectious diseases in interstitial areas. Platelets result from cell fragmentation and are involved with clotting. Platelets are cell fragments that bud off megakaryocytes in bone marrow. They carry chemicals essential to blood clotting. Platelets survive for 10 days before being removed by the liver and spleen.

There are 150,000 to 300,000 platelets in each millilitre of blood. Platelets stick and adhere to tears in blood vessels; they also release clotting factors. A hemophiliac's blood cannot clot. Providing correct proteins (clotting factors) has been a common method of treating hemophiliacs. It has also led to HIV transmission due to the use of transfusions and use of contaminated blood products.

Lymphatic System: Water and plasma are forced from the capillaries into intracellular spaces. This interstitial fluid transports materials between cells. Most of this fluid is collected in the capillaries of a secondary circulatory system, the lymphatic system. Fluid in this system is known as lymph. Lymph flows from small lymph capillaries into lymph vessels that are similar to veins in having valves that prevent backflow. Lymph vessels connect to lymph nodes, lymph organs, or to the cardiovascular system at the thoracic duct and right lymphatic duct. Lymph nodes are small irregularly shaped masses through which lymph vessels flow. Clusters of nodes occur in the armpits, groin, and neck. Cells of the immune system line channels through the nodes and attack bacteria and viruses travelling in the lymph.

7

Regulation of Flowering (Plants)

Plants must often gain competence to respond to internal or external signals regulating flowering. Once, plants are competent to reproduce, a combination of factors including light, temperature, and both promotive and inhibitory internal signals determine when a flower is produced. These signals turn on genes that specify where the floral organs, sepals, petals, stamens, and carpels will form.

The decision of a plant to initiate flowering often involves coordinating the perception of environmental cues such as day length, light conditions and temperature with endogenous factors such as plant developmental status and age. This coordination ensures that the transition is properly timed to achieve maximal reproductive success.

In Arabidopsis, several components of flowering pathways that have been defined by genetics and physiology are involved in modifying the chromatin of target genes. Arabidopsis flowering is regulated by four major pathways.

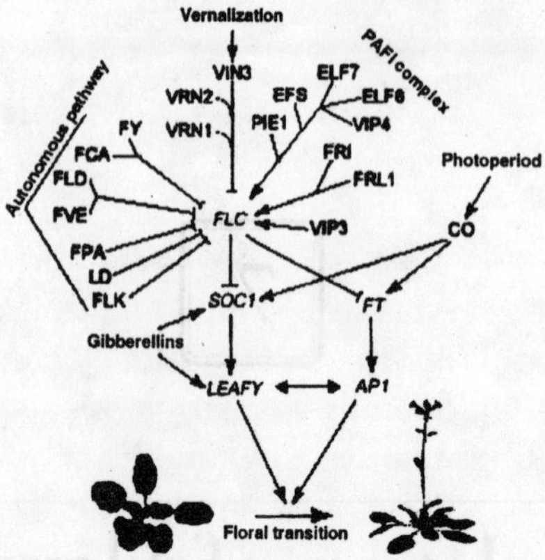

Pathways regulating flowering time in Arabidopsis.

Arabidopsis is a facultative long-day plant and the photoperiod pathway accelerates flowering in response to increasing day length. The vernalisation pathway renders Arabidopsis competent to flower after longterm cold exposure, the autonomous pathway constitutively represses flowering and gibberellins promote flowering and are required for flowering in non-inductive photoperiods. Inputs from these pathways are integrated to control the expression of a set of common downstream target genes such as Suppressor of overexpression of constant 1 (SOC1) and FT, which are called 'flowering-time integrators'. SOC1 and FT in turn activate floralmeristem-identity genes such as Leafy and Apetala 1, which cause the production of flower primordia. There is natural flowering-time variation among Arabidopsis accessions.

Vernalisation

The winter-annual (late flowering without vernalisation) versus rapid-cycling (early flowering) habit is determined by allelic variation at FRIGIDA (FRI) and Flowering Locus C (FLC). Winter annuals have dominant alleles of FRI and FLC, whereas rapid-cycling types have either a non-functional fri allele or a weak flc allele. FLC is a MADS-box transcription factor that quantitatively blocks the floral transition, in part by repressing expression of the flowering-time integrators SOC1 and FT. FRI elevates FLC expression to levels that inhibit flowering. Vernalisation represses the FRI-mediated increase in FLC expression.

The photoperiod and autonomous pathways were identified in screens for late-flowering mutants in mutagenised rapid-cycling backgrounds [e.g. Columbia and Landsberg erecta (Ler)]. The role of autonomous pathway genes in flowering is to repress FLC expression. Many summer annuals are early flowering without vernalisation because, in the absence of a functional FRI allele, the autonomous pathway keeps FLC levels low.

In winter annuals, FRI overcomes the autonomous-pathway-mediated repression of FLC. Therefore, FLC regulation is a convergence point for the control of flowering time by vernalisation, FRI and the autonomous pathway. The regulatory mechanisms by which FLC expression is controlled are of great interest given its central role in controlling flowering time.

Recent screening for mutants that render winter-annual, FRI-containing lines early flowering have led to the identification of genes that are additional positive regulators of FLC expression. Some of these are required for the elevated FLC expression in both autonomous-pathway mutants and FRI-containing lines {e.g. Photoperiod-Independent Early Flowering 1 (PIE1), Vernalisation Independence 3 (VIP3) and VIP4), whereas others are specifically required only for. FRImediated FLC activation (e.g. Fri Like 1).

The recent characterisation of FLC repressors and activators has shown that some of these regulatory proteins are involved in the covalent modification of FLC chromatin. Chromatin modifications such as the acetylation and methylation of specific residues of specific histones are typically referred to as the 'histone code'. These chromatin modifications provide an additional level of regulation of gene expression beyond that of the transcription factors that recruit RNA polymerase to target genes. Recent studies of FLC regulation provide perhaps the best-studied example in plants of how changes in the chromatin environment are used to modify a developmental programme.

Flowering Locus C Expression

Activation

H3-K4 Hyper-trimethylation of FLC Chromatin is Associated with the delay of Flowering in Winter-annual Arabidopsis: Trimethylation of histone 3 (H3) at lysine 4 (H3-K4) is associated with the chromatin of active genes. Lysine-4 methylation of H3 occurs in three states: mono-, di- and trimethyl.

In yeast and Drosophila, the presence of a trimethylated K4 defines an active state of gene expression. Other reports indicate that dimethylation of H3-K4 might also associate with active genes in multicellular eukaryotic organisms. Recent studies in yeast have provided an outline of how H3-K4 trimethylation levels are increased and how this chromatin modification leads to gene activation. The yeast RNA Polymerase II Associated Factor 1 (PAF1) complex associates with the RNA polymerase II complex during transcription.

One role of the yeast PAF1 complex is to recruit a H3-K4 methyltransferase known as SET1 to target genes. SET1 generates H3-K4 trimethylation predominantly in the 50 portion of actively transcribed regions. The resulting increase in H3-K4 trimethylation links transcription and chromatin structure, and is thought to serve as a 'transcriptional memory' that establishes a positive feedback loop of gene activity.

Screens for mutants in which winter annuals are converted into rapid-flowering types because of attenuation of FLC expression have led to the identification of several Arabidopsis relatives of the yeast PAF1 complex. The yeast PAF1 complex consists of PAF1, CTR9, LEO1, CDC73 and RTF1.

The Arabidopsis relatives of PAF1, CTR9 and LEO1 are present in the genome as single copies. Lesions in Early Flowering 7 (ELF7, the relative of yeast PAF1), Early Flowering 8 (ELF8, the relative of yeast CTR9) or Vernalisation Independence 4 (VIP4, the relative of yeast LEO1) suppress the ability of FRI to increase FLC expression and thus to convert winter annuals into rapid-flowering types.

When tested, these lesions also suppress the increased FLC expression in autonomous-pathway mutants, indicating that the Arabidopsis PAF1 complex is generally required for high levels of FLC expression. In a FRI-containing winter annual line (in which FLC is highly expressed), FLC chromatin has increased levels of H3-K4 trimethylation compared with a rapid-flowering line that lacks an active FRI allele and has relatively low levels of FLC expression.

Late-flowering autonomous pathway mutants also have increased levels of H3-K4 trimethylation compared with the rapid-flowering parental line. Thus, FRI is formally a promoter of H3-K4 trimethylation of FLC chromatin, and autonomous pathway genes are repressors of this chromatin modification. The increased levels of H3-K4 trimethylation are present in a distinct region of FLC chromatin around the transcription start site, the first exon and a 50 portion of the first intron.

In elf 7 and elf 8 mutants, the H3-K4 trimethylation of FLC chromatin is greatly reduced. These studies indicate that the PAF1-like complex is necessary for the increased FLC expression of FRI containing lines or autonomous-pathway mutants by increasing H3-K4 trimethylation of FLC chromatin. The elf7 and elf8 mutants also cause early flowering in non-inductive short days. This is due, at least in part, to a requirement for the Arabidopsis relatives of the PAF1 complex components for expression of additional MADSbox genes.

In Arabidopsis, there are five FLC relatives: Flowering Locus M (FLM), Mads Affecting Flowering 2 (MAF2), MAF3, MAF4 and MAF5. Like FLC, FLM and MAF2 are also floral repressors. However, unlike FLC, FLM and, possibly, MAF2 repress flowering in non-inductive photoperiods. In elf7 and elf8 mutants, FLM and MAF2 expression is greatly reduced, and FLM chromatin has reduced levels of H3-K4 trimethylation.

Thus, the Arabidopsis PAFl-like complex might be required for the expression of the FLC clade of MADS-box genes, and loss of expression of this clade would lead to both early flowering in non-inductive short days and a conversion of winter annuals into rapid-flowering types in inductive long days. SET1 is the only H3-K4 methyltransferase in yeast.

During the course of evolution, plants have undergone extensive duplication of SET-domain proteins. Arabidopsis has several potential H3-K4 methyl transferases, based on sequence and domain similarities to known K4 methyl transferases in other organisms. Recent work (S.D. Michaels et al., unpublished) has revealed that a particular Arabidopsis relative of yeast SET1 and the Drosophila H3-K4 methyltransferase ASH1 is necessary for H3-K4 trimethylation in FLC chromatin.

Lesions in the SET-domain gene early flowering in short days (EFS; this locus has been genetically characterised before), like those in ELF7 and ELF8, cause early flowering in short days as well as suppression of FRI-mediated late flowering and reduction in trimethylation levels of FLC chromatin and suppression FLC expression. It thus appears that, similar to the situation in yeast, the Arabidopsis PAF1-like complex recruits EFS to methylate target genes.

Connection between H3-K4 Trimethylation and ATP Dependent Nucleosome Remodelling: Recently, it has been shown that ISW1p (a yeast ATP hydrolysing, chromatin-remodelling protein) preferentially binds di- and trimethylated H3-K4 and H3-K4 methylation of chromatin of certain actively transcribed genes is required for association with ISW1p in vivo. An Arabidopsis relative of ISW1p, PIE1, has been found to be necessary for FLC expression.

Also, the range of phenotypes of the pie1 mutant, such as early flowering in short days and a suppression of FRI-mediated late flowering, is similar to that of efs, elf7 and elf8 mutants. Thus, it is tempting to speculate that PIE1 binds trimethylated K4 and remodels chromatin of FLC and its relatives to increase gene expression. Overall, the studies of the regulation of flowering time and FLC expression indicate that plants and yeast use H3-K4 trimethylation in similar ways.

Repression

Certain Autonomous-pathway Components Repress FLC Expression by Histone Deacetylation: The autonomous pathway consists of seven known genes: FCA, Flowering locus D (FLD), FPA, FVE, FY, Luminidependens (LD) and flowering locus K. The autonomous pathway was originally defined by late-flowering mutations that retained a photoperiod response.

Subsequent molecular and genetic studies have defined the autonomous pathway as repressors of FLC. Histone acetylation is another chromatin modification that affects gene expression. Generally, hyperacetylation of core histone tails of H3 and H4 is linked to actively expressed genes, whereas hypoacetylation of these histone tails is associated with inactive genes.

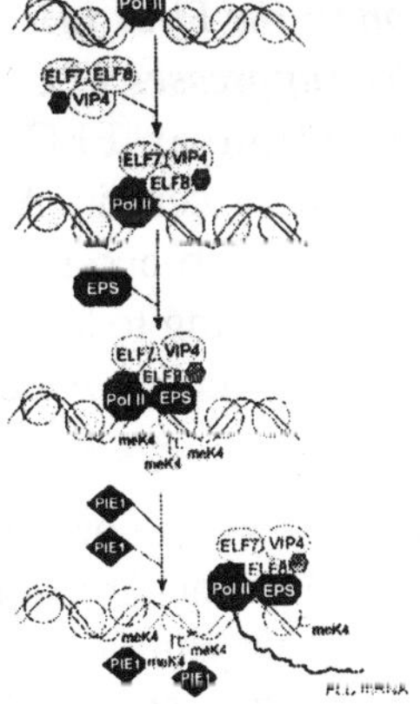

FLC locus without vernalisation

The recent cloning of two autonomous-pathway genes, FLD and FVE, revealed that these genes encode proteins similar to components' of a mammalian histone deacetylase (HDAC) complex: FLD and FVE are relatives of human KIAA0601 and RbAp48/46, respectively. Indeed, histone acetylation levels of FLC chromatin were increased in fld and fve mutants but not in other autonomous mutants including fca, fpa and ld.

This indicates that FLD and FVE participate in the deacetylation of FLC chromatin as components of an HDAC complex. Furthermore, the lack of an effect on the histone acetylation level of FLC chromatin in other autonomous pathway mutants indicates that the autonomous pathway is not a typical linear pathway, but must be composed of at least two sub-pathways that repress FLC expression by different mechanisms.

This molecular subdivision of the autonomous pathway is consistent with the previous genetic studies that indicated tha: subpathways exist. Recent studies also indicate that a cis-regulatory region in the first intron of FLC is necessary for deacetylation. However, the protein(s) that binds to this regulatory region and recruits the FLD-containing HDAC complex to FLC chromatin has not yet been discovered.

H3-K9 Methylation in the First Intron of FLC Directed by Short Interfering RNAs Suppresses FLC Expression: Naturally occurring allelic variation at FLC contributes to the rapid-flowering behaviour of certain Arabidopsis accessions such as Ler and Da (1)-12. Both accessions have a functional FRI but are early flowering owing to lower steady-state mRNA levels of FLC; that is, these alleles of FLC are not effectively upregulated by FRI.

In Ler, there is a 1.2-kb insertion of a Mutator-like transposable element (TE) in the first intron of FLC. This TE is clearly responsible for the inability of FRI to upregulate FLC because transfer of this TE to a strong FLC allele causes it to behave like the Ler allele. This TE insertion renders Ler FLC subject to repressive chromatin modifications mediated by silencing short interfering RNAs (siRNAs) generated from related TEs elsewhere in the genome. Specifically, the level of H3-K9 dimethylation in a region of Ler FLC chromatin including the TE insertion and an intronic region immediately downstream of TE is elevated.

Although methylation of K4 of H3 is associated with active genes, other lysine methylations of H3 (such as methylation of K9 and K27) are characteristic of inactive genes that adopt a repressive, heterochromatin like conformation. Furthermore, siRNAs are known to direct H3-K9 dimethylation to regions of sequence relatedness. Thus, it is likely that siRNAs trigger histone H3-K9 dimethylation, resulting in a heterochromatin 'island' in the middle of FLC chromatin, which causes transcriptional gene suppression in Ler FLC.

The Da (1)-12 ecotype has a different class (Copialike) of TE in its first FLC intron and, similar to the situation in Ler FLC, there is an island of heterochromatin around the TE in the Da (1)-12 FLC allele. Thus, the conversion of a winter annual into a rapid-flowering type by TE insertions in FLC provides another example of McClintock's model that insertion of a TE can result in 'the transposition of heterochromatin'. These examples of TEs in Ler FLC and Da FLC result in a large effect on flowering time that might have had adaptive value.

Epigenetic Memory of Winter in Arabidopsis is a Result of FLC Chromatin Modifications: Vernalisation-requiring plants such as winter annuals and biennials require exposure to the prolonged cold of a typical winter to acquire the competence to flower. In Arabidopsis, vernalisation establishes this competence largely by repressing FLC expression. Cold-mediated FLC repression is stably maintained after plants resume growth in warm conditions.

The mitotic stability of the vernalised state in Arabidopsis in the absence of the inducing signal (cold exposure) is characteristic of an epigenetic switch, and this switch permits plants to 'remember' winter at a cellular level. This memory is not passed to the next generation; loss of this memory is necessary so that each generation will have a vernalisation requirement.

Screening for mutants that are not able to become vernalised after a long-term cold exposure has led to identification of three genes involved in vernalisation: Vernalisation 1 (VRN1), VRN2 and Vernalisation insensitive 3 (VIN3). Recent studies have shown that vernalisation leads to a series of repressive modifications in FLC chromatin, which converts active FLC chromatin into a heterochromatin-like state.

During vernalisation, histone tails of FLC chromatin are deacetylated, followed by an increase in H3-K27 and H3-K9 methylation. VIN3, a plant-homeodomain-containing protein, is likely to play a role in the initiation of these modifications because none of these vernalisation-mediated modifications are observed in FLC chromatin of vernalised vin3 mutants. VRN2 is a plant relative of the Drosophila Polycomb-group protein Suppressor of Zeste-12, and is thought to be a component of a Polycomb repressor complex that mediates H3-K27 methylation of FLC chromatin, which probably leads to H3-K9 methylation.

VRN1, a B3-domain containing DNA-binding protein, is required for H3-K9 methylation of FLC chromatin during vernalisation. Vernalisation is dominant over the FLC activation caused by FRI and FRL1 or by lesions in autonomous pathway genes. Recent studies show that, like lesions in ELF7 and ELF8, vernalisation causes a reduction of H3-K4 trimethylation in FLC chromatin in FRI-containing lines.

The reduction of H3-K4 trimethylation does not appear to be the result of vernalisation suppressing the expression of the genes involved in this modification of FLC chromatin; expression of ELF7, ELF8, VIP4 or EFS is not affected by vernalisation. One of several possibilities to account for the vernalisation-mediated decrease of H3-K4 trimethylation is that after vernalisation the PAF1 complex is unable to access FLC chromatin owing to its heterochromatin-like nature.

8

Commercial Significance (Plants)

Our lives are enriched by beautiful hardwoods, spices, essential oils and fruits. In addition, tropical countries export many fibres, gums, resins, dyes, and plant essences that we may never see directly, but which are widely used in medicine and industry. For instance, Latin America and Africa are major suppliers of coffee and cacao (from which we derive chocolate), while Asia produces most of our rice and natural rubber.

Plant products like those just mentioned are often referred to as "commodities" or "cash crops." Unlike many exports from the industrialised economies, commodities are usually exported in minimally processed states as raw materials. Whether tropical nations should continue to rely extensively on these exports to fuel their emerging economies is a hotly debated subject, with critics maintaining that overproduction depresses world prices of these materials and diverts arable land from food production for local markets. Regardless, patterns of trade in commodities are not likely to change significantly in the near future.

As tropical nations seek to increase their share in the world marketplace, a key question is the best way to balance these strategies with the needs to conserve and manage remaining forested areas. Indiscriminate harvesting techniques and clearing large tracts for cultivation or ranching have been all too characteristic of the past. The future will require more appropriate means of extracting plants or their products if we are not to lose the many thousands of other tropical species holding genetic "blueprints" important to our future. This will require strong international leadership on economic and environmental fronts and, for all citizens of the world, a willingness to rethink our use of the Earth's resources.

Bamboo

Bamboo is the world's largest grass and is native to many parts of the world. It is grown throughout the tropics and is used as a source of food as well as for construction and weaving. It has great strength and is used in Japan as scaffolding for building skyscrapers.

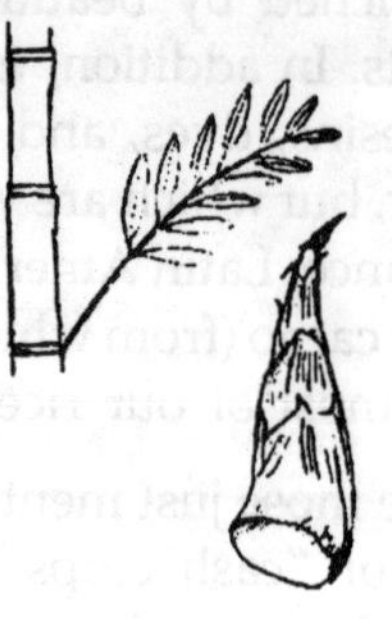

Bamboo

Banana and Plantain

Bananas and plantains are very closely related. Plantains, however, are used more like a starch than a fruit. Both are native to East Asia and Australia but are now grown throughout the tropics. Though commonly called "trees", they are actually the world's largest herbaceous plants, as they do not develop woody stems like trees.

Banana

Sugar Cane

The stems of this grass plant can reach 10 feet tall and are rich in sugar. The stems are crushed to extract the sweet juice, which is then processed to obtain sugar. Sugar cane is native to New Guinea. It was introduced into the New World by Columbus on his second voyage in 1493.

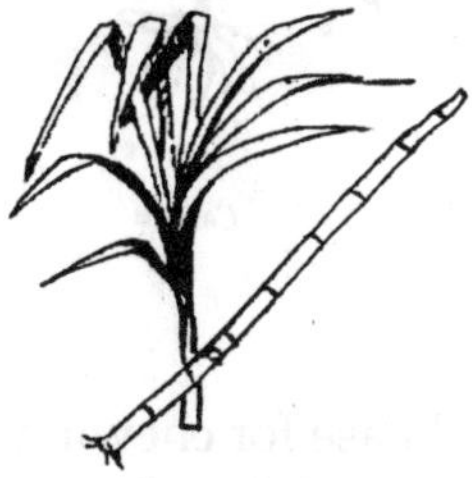

Sugar Cane

Nutmeg and Mace

Nutmeg and mace come from different parts of the same fruit. Nutmeg is the seed and mace is the red net-like fibre (aril) that surrounds the seed. Nutmeg is native to the Moluccas (Spice Islands).

Nutmeg and Mace

Cassava

Native to Mexico, Guatemala, and northern Brazil, cassava is now grown throughout the tropics. It is the starch staple of over 500 million people. The roots are peeled and boiled (like potatoes) or made into a flatbread. It is in the same family as poinsettia and wild varieties can be toxic if not prepared correctly. Tapioca is made from heated, purified cassava starch and is a common thickening agent.

Cassava

Chicle or Sapodilla

Chicle, the original base for chewing gum, comes from the latex (sap) of a tree native from southern Mexico to northern Brazil. When the trunks are scored with diagonal down-sloping cuts, the latex flows from the bark. It is collected, molded into blocks, and shipped for processing, where sugar and flavouring are added. Today chicle has been largely replaced with the latex from other trees and by synthetic gums. The fruit of the chicle tree, sapodilla, is delicious.

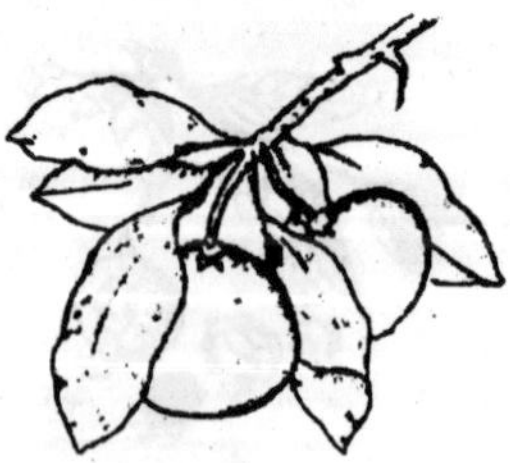

Chicle

Allspice

Allspice is not a mixture of spices but rather a single fruit that contains the flavours of nutmeg, cinnamon and cloves. The Maya used allspice to embalm the bodies of their leaders.

Allspice

Rubber

Rubber trees are native to the Brazilian Amazon but now are also grown in plantations in South East Asia. Diagonal slashes are made in the bark of the tree and the latex sap that exudes is collected. The latex is mixed with water and heated over a smoky fire to produce a ball of rubber ready for processing. Today most rubber is synthetic; natural rubber is restricted to specialised uses.

Rubber

Cacao

Cacao is native to the eastern Andes. The Maya and Aztecs made a beverage from the seeds of this plant calling it the "food of the gods." The seeds, after being fermented and pulverised, were mixed with water to produce a beverage. It was not until Europeans added sugar and milk that the world came to know chocolate as we do today. Today, major production areas are in West Africa, Brazil, and Mexico.

Cacao

Vanilla

Vanilla is the only orchid that is grown for purposes other than its flowers. It is native to Mexico, and was introduced into Europe in the mid-16th century by the Spanish. The flowers are pollinated by hand or by a small bee and produce an elongated fruit, or "bean", which is fermented to produce vanilla beans. True vanilla flavouring is extracted from the fermented seed pods with alcohol. Today most of the world's supply is grown in the Seychelles, Central America, and Madagascar.

Vanilla

Ginger

Native to southern Asia, ginger is now grown throughout the tropics with major production in Nigeria, Sierra Leone, Jamaica, and India. The plant's rhizome (stem that grows at ground level like iris) is used fresh as well as dried and ground. It is a common ingredient in Asian cooking and flavours ginger snaps, ginger ale, ginger beer, ginger cake, and pumpkin pie.

Ginger

Piper Nigrum

Black pepper and hot pepper (Capsicum) are not closely related. Black pepper originated in India and Cambodia and was the spice that, above all, drove the spice trade. The green, immature fruits of the pepper vine are harvested and dried, turning black upon drying. White pepper is obtained by letting the "berries" turn red (ripen) and then removing the outer husk, leaving the straw-coloured "kernel". It is milder than black pepper.

Black pepper

Jackfruit

Native to South East Asia, jackfruit is now grown throughout the tropics. The fruits can weigh up to 100 pounds and are produced directly on the trunks and branches. Fruits are green or yellow and "warty" on the surface. Inside, the seeds are surrounded with a fragrant yellow flesh tasting like pineapple and bananas. The seeds are also eaten boiled or roasted.

Jackfruit

Rattan

Native to Asia, rattan is the collective name for the climbing palms. They are rope-like, woody climbing plants that can grow to 600 feet long. After being collected from the forest the stems are boiled in oil and scoured in sand or sawdust to remove their natural gums and resins. The outer skins are removed and used to weave furniture, baskets, screens, and chair seats. The inner core is used for making baskets.

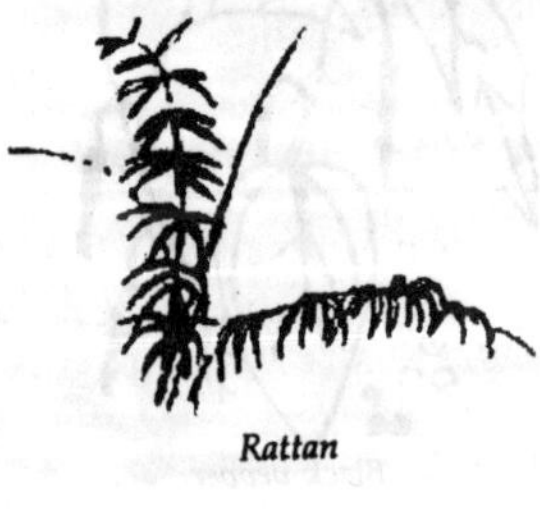

Rattan

Cola or Kola

Cola is related to cacao, the plant from which chocolate is made, and is best known as a flavouring in cola soda pops. Cola seeds, called "nuts", are rich in caffeine. It is native to West Africa but now is also grown commercially in Jamaica and Brazil. In West Africa a cola drink is made by mixing the pulverised, fermented and dried beans with water.

Cola

Ebony

Ebony is in the same genus as the North American persimmon, D. virginiana, and has a similar fruit. Its heartwood gives it commercial value. The wood is extremely hard and will sink in water. It is used for piano keys, cutlery handles, musical instruments and carvings. It is native to West Africa. Good quality logs are rare due to over harvesting.

Ebony

Jojoba

Jojoba (Simmondsia chinensis) is an excellent desert crop. This evergreen woody shrub thrives well in coastal non-saline wastelands and sand dunes under semi-arid conditions as well as on rocky soils without water logging. It withstands temperature upto 500C and grazing by cattle. The plant requires a small quantity of water during initial 3 yr. Thereafter a minimum life saving irrigation is needed. Jojoba absorbs carbon monoxide and dust from atmosphere and helps in controlling pollution and soil erosion.

The seeds are processed to yield a valuable industrial oil (45-60 per cent in seed). The oil does not turn rancid during storage upto 25 year and does not degrade at temperature upto 3000C. It is used as lubricant for high -speed high-temperature machinery, as coating on pharmaceutical tablets, as base in cosmetic industry and as hair restorer.

The plant yields 350 kg seeds per ha during initial 3-10 period and commercial yields of 500-800 kg are achieved thereafter upto its shelf life of 100-150 year.

Salvadora Persica

Salvadora persica is a evergreen drought-tolerant tree that grows very well on coastal sand duns, marginal to high saline wastelands with or without water logging, ravines and saline/ alkaline dry zones. During initial 3 year, it requires some irrigation to establish to the plant. Subsequently minimum life saving irrigation is needed. It can also be irrigated with saline water. The plants can yield 3.5t seeds/ha/year. The plant has a life upto 100 year.

Asparagus Officinalis

Asparagus officinalis is mainly grown for edible vegetable purpose in Germany, USA, Spain & France. Asparagus comes in to production after 3rd year of cultivation and continues to give crop upto 15year. An improved variety, viz. SL-831 yields

3.3 t/ha. The cost benefit ratio is 1:1:2 in established period and 1:4:3 in subsequent 7 year.

Piper Beetle

Beetle leaves are used as a masticatory alongwith arecanut, lime and catechu. This perennial climber is cultivated in closed conservatories. The yield is nearly 10t/ha/year.

There are mainly five cultivars of betelvine, viz. Kappori, Desawari, Bangia, Meetha and Sanchi. While Kappori and Sanchi are the principal cultivars in the peninsular India, Bangia and Desawari are common in north India. Cultivar, Meetha is grown on commercial scale in West Bengal only. Betelvine is cultivated over an area of 40,000 ha in the country. It is a capital and labour intensive cash crop.

With technical inputs and improved agrotechniques, it has been possible to improve betelvine cultivation, increase yield per unit area, extend storage life and improve the quality of leaves.

Ornamental Plants

The varied climatic conditions and soils of India are congenial for cultivation of almost all types of ornamental plants in one part or the other. The R&D effects on ornamentals have resulted in the evolution of elite cultivars and development of improved agrotechnologies.

Nearly 200 new cultivars have been evolved for 15 well known ornamental plants, viz., Amaranthus, Antirrhinum, Bougainvillea, Chrysanthemum, Erythrina, Gladiolus, Hibiscus, Hippeastrum, Lantana, Nyctanthes arbor-tristris, Polianthes tuberosa, perennial Porttulaca, Rosa, Tagetes and Verbena. The cultivation of ornamentals for nursery and cut flower trade is a highly remunerative.

Gladiolus

Gladiolus is an important floral crop for the cut-flower trade. Its magnificent spikes in wide range of colours remain fresh for a pretty long time. Cut spikes are much valued for making floral arrangement for interior decoration and are in constant demand. The cultivation was earlier confined mainly to the temperate and mild climate regions of the country.

An appropriate technology for the production of quality gladiolus spikes and corms in north Indian plains has been standardised. The planting material, i. e. corms and corm lets of gladiolus mostly American and Dutch hybrids, can be made available.

Annatto

Annatto is a natural carotenoid obtained from the seeds of Bixa orellana and is used for colouring cheese, butter, margarine, fruit juices sausage, candles, etc.

Sub-tropical climate with humid atmosphere is suitable for the cultivation of annatto seeds. The optimum temperature for growth is 150C-350C, rainfall approx. 200cm and well drained soil. Annatto can be propagated through seeds. Seedlings are raised during March-May in the seedbeds and plantation done in June-July. The plant bears fruit after 2 year. There are 12 to 18 pericarps in a bunch and 24 to 40 seeds in a pericarp. When the first pod opens out, the whole bunch along seeds are taken out by beating with sticks. A plant yields 500g-l kg seeds per annum. It is a highly remunerative crop.

Saponin Bearing Plants

The commonly known soapnut trees, (Sapindus emarginatus and Sapindus mukorossi) are widely distributed in Karnataka, Andhra Pradesh, Orissa, Himachal Pradesh and other northern parts of India. The detergent property in soapnut is due to the presence of triterpene saponins. The saponin content in soapnuts varies from 10 to 18 per cent. The saponins are used in detergents, shampoos, textiles, foods, pharmaceuticals and photo film industry. The photo film industry alone requires about 6t of saponin per annum, presently being totally imported.

The process has been developed for isolation and purification of saponins from the soapnuts. The soapnuts are cut and separated into pericarp and the seeds. The pericarp is used in the saponin. Technologies are also available for integrated utilisation of sapolin on cottage scale for use in shampoos & emulsifiers, collection and evaluation of honey from Sapindus flowers as well as utilisation of seeds for oil, soap, seeds cake, etc.

Fast Growing Species

The fast growing plants have assumed increasing importance due to the depletion of green cover. These provide fodder, fuel, timber, biomass for paper, pulp and other valuable industrial products. These can be also be adopted in water logged areas, marshes, swamps, Jhum lands, sandy, clayey soils, gullies, ravines, denuded and eroded soils, mined wastelands, etc. The plants possess fast rate of growth, need less input and improve soil and environment.

Economic Seaweeds

Agarophyte seaweeds, viz Gracilaria edulis and Gelidiella acerosa are used for the production of agar. There is great potential for the cultivation of this material because of its non- availability from the wild stock. Cultivated material is pure and fetches higher sheltered calm water. The cultivation is done by vegetable propagation starting from fragments.

The Single Rope Floating Raft (SPFR-) technique has been developed which is suitable for a wide area and greater depth. Substrate is cheep and long polypropylene rope of 10 mm dia. It is attached to 2 wooden stakes with 2 synthetic fibre anchor cables and kept afloat with synthetic floats. The length of the cable is twice the depth of the sea (3-4m). Each raft is kept afloat by means of 25-30 floats. The cultivation rope is hung with the floating rope. A stone is attached to the lower end of the cultivation rope to keep it in vertical position. Generally 10 fragments of Gracilaria edulis are inserted in each rope. The distance between two rafts is kept at 2m.

The cultivation of seaweeds can be beneficial to the weaker sections, women, fishermen through generation of income and employment.

9

Cell Structure

Cell is the smallest unit of life. The cell is the starting point for all organisms. Some living things (such as bacteria) are composed of one cell, whereas a complex organism such as the human body is composed of trillions of cells. In the multicellular animals, a group of cells makes up tissues, which make up organs, which make organ systems, which in turn form the complete animal. Without cells we would not exist.

There are three basic components of every cell: an outer membrane, a central nuclear region, and the cytoplasm in between. These elements don't make much sense if you don't know why they are important to the cell. The cell has a membrane. The membrane surrounds the cell, isolating it from the outside. The membrane is complex and can contain many channels that allow the cell to communicate with the environment through complicated chemical interactions that happen on the scale of a few molecules at a time. The membrane also regulates the in-and-out flow of certain materials, allowing certain chemicals in (such as food) or out (such as waste), but blocking others.

Even the simplest cells have a nuclear region. Think of this region as a control centre. The nuclear region is the social director for the cell. It basically tells the cell what to do and when to do it. These orders are handed down by the nucleus in the form of chemical reactions that take place in the cell. The cytoplasm is not something from Ghostbusters, but instead it is a semi-fluid matrix which occupies the volume between the nuclear region and the cell membrane. Think of the cytoplasm as the cream filling. The cytoplasm, however, has a more important function: it contains the chemical wealth of the cell. The sugars, amino acids, and proteins that are used to carry out the chemical reactions of the cell are housed within the cytoplasm. All cells share this basic architecture.

Not all cells are the same size, but for the most part they are smaller than you can imagine. The reason that cells are so small is that each cell must maintain centralised control to function efficiently. If the cell was large it would take a long time to perform the tasks given by the nucleus, since the chemical reactions involved in a cell's function are fairly complicated and do not work well on a larger scale. Smaller cells also have the advantage of a greater surface area for their mass, thus allowing them to communicate with their environment. The small size of cells did not allow them to be discovered until 1665, when the first microscope was invented.

What kind of cells are there? For the most part there are Prokaryote cells and Eukaryote cells. An easy way to remember the difference is thinking of "P" standing for "prehistoric." Prokaryotes are the more basic of the two kinds of cells. A prokaryote cell is very simple in its structure and function. Bacteria are prokaryote cells. These cells have a membrane and a very simple nuclear region. There are not separate divisions in the bacterium that carry on their own chemical reactions. For the most part, in a bacteria cell, what you see is what you get.

The Eukaryote cell is the more advanced of the two, and most cells are eukaryotic in nature. The structure of these cells is much more complicated and diverse than the structure of prokaryotic cells. The most distinctive difference is that the eukaryotic cell has extensive membrane subdivisions called organelles, which are kind of like the individual organs in a human body. Each organelle of the eukaryote has its own function. For example, eukaryotic cells have lysosomes which contain digestive enzymes that play a role in cell death.

Some cancer research has been linked to lysosomes, in that a gene for cancer may temporarily disable the lysosome from doing its cellular duty (i.e., killing off the cell when it is time). Thus the cancer cells divide and these bad cells never get weeded out of the body. It is thought that eukaryote cells may have evolved from the simpler prokaryote cells. Cellular biology is one facet of biology that is intensely studied. Cells are very important because they carry out all of life's activities. Without cells, we would not even be able to move a muscle, because the reactions that are needed to allow it to happen would not be present.

Cell Theory

The Cell theory, or cell doctrine, states that all organisms are composed of similar units of organisation, called cells. The concept was formally articulated in 1839 by Schleiden & Schwann and has remained as the foundation of modern biology. The idea predates other great paradigms of biology including Darwin's theory of evolution, Mendel's laws of inheritance, and the establishment of comparative biochemistry. Ultra-structural research and modern molecular biology have added many tenets to the cell theory, but it remains as the preeminent theory of biology.

In 1838, Theodor Schwann and Matthias Schleiden were enjoying after-dinner coffee and talking about their studies on cells. It has been suggested that when Schwann heard Schleiden describe plant cells with nuclei, he was struck by the similarity of these plant cells to cells he had observed in animal tissues. The two scientists went immediately to Schwann's lab to look at his slides. Schwann published his book on animal and plant cells the next year, a treatise devoid of acknowledgments of anyone else's contribution, including that of Schleiden.

He summarised his observations into three conclusions about cells:

(1) The cell is the unit of structure, physiology, and organisation in living things.
(2) The cell retains a dual existence as a distinct entity and a building block in the construction of organisms.
(3) Cells form by free-cell formation, similar to the formation of crystals (spontaneous generation).

We know today that the first two tenets are correct, but the third is clearly wrong. The correct interpretation of cell formation by division was finally promoted by others and formally enunciated in Rudolph Virchow's powerful dictum, "Omnis cellula e cellula"...

"All cells only arise from pre-existing cells". The modern tenets of the Cell Theory include:

1. All known living things are made up of cells.
2. The cell is structural & functional unit of all living things.
3. All cells come from pre-existing cells by division. (Spontaneous Generation does not occur).
4. Cells contains hereditary information which is passed from cell to cell during cell division.
5. All cells are basically the same in chemical composition.
6. All energy flow (metabolism & biochemistry) of life occurs within cells.

As with any theory, its tenets are based upon previous observations and facts, which are synthesised into a coherent whole via the scientific method. The Cell Theory is no different being founded upon the observations of many. Credit for the first compound (more than one lens) microscope is usually given to Zacharias Jansen, of Middleburg, Holland, around the year 1595. Since Jansen was very young at that time, it's possible that his father Hans made the first one, but young Jansen perfected the production. Details about the first Jansen microscopes are not clear, but there is some evidence which allows us to make some guesses about them (Jansen microscopes).

In 1663 an English scientist, Robert Hooke, discovered cells in a piece of cork, which he examined under his primitive microscope. Actually, Hooke only observed cell walls because cork cells are dead and without cytoplasmic contents. Hooke drew the cells he saw and also coined the word CELL. The word cell is derived from the Latin word 'cellula' which means small compartment. Hooke published his findings in his famous work, Micrographia: Physiological Descriptions of Minute Bodies made by Magnifying Glasses. Ten years later Anton van Leeuwenhoek, a Dutch businessman and a contemporary of Hooke used his own (single lens) monocular microscopes and was the first person to observe bacteria and protozoa.

Leeuwenhoek is known to have made over 500 "microscopes," of which fewer than ten have survived to the present day. In basic design, probably all of Leeuwenhoek's instruments were simply powerful magnifying glasses, not compound microscopes of the type used today. Leeuwenhoek's skill at grinding lenses, together with his naturally acute eyesight and great care in adjusting the lighting where he worked, enabled him to build microscopes that magnified over 200 times, with clearer and brighter images than any of his colleagues at that time. In 1673, Leeuwenhoek began writing letters to the newly formed Royal Society of London, describing what he had seen with his lenses. His first letter contained some observations on the stings of bees.

For the next fifty years he corresponded with the Royal Society. His observations, written in Dutch, were translated into English or Latin and printed in the Philosophical Transactions of the Royal Society. Leeuwenhoek looked at animal and plant tissues, at mineral crystals, and at fossils. He was the first to see microscopic single celled protists with shells, the foraminifera, which he described as "little cockles... no bigger than a coarse sand-grain." He discovered blood cells, and was the first to see living sperm cells of animals. He discovered microscopic animals such as nematodes (round worms) and rotifers. The list of his discoveries is long. Leeuwenhoek soon became famous as his letters were published and translated.

Between 1680 and the early 1800's it appears that not much was accomplished in the study of cell structure. This may be due to the lack of quality lens for microscopes and the dedication to spend long hours of detailed observation over what microscopes existed at that time. Leeuwenhoek did not record his methodology for grinding quality lenses and thus microscopy suffered for over 100 years.

Around 1833 Robert Brown reported the discovery of the nucleus. Brown was a naturalist who visited the "colonies of Australia" from 1801 through 1805, where he catalogued and described over 1,700 new species of plants. Brown was an accomplished technician and an extraordinarily gifted observer of microscopic phenomena. It was Brown who identified the naked ovule in the gymnospermae. This is a difficult observation to make even with a modern instrument and the benefit of hindsight. But it was with the observation of the incessant agitation of minute suspended particles that Brown's name became inextricably linked.

The effect, since described as Brownian Movement, was first noticed by him in 1827. Having worked on the ovum, it was natural to direct attention to the structure of pollen and its Brown interrelationship with the pistil. In the course of his microscopic studies of the epidermis of orchids, discovered in these cells "an opaque spot," which he named the nucleus. Doubtless the same "spot" had been seen often enough before by other observers, but Brown was the first to recognise it as a component part of the vegetable cell and to give it a name. This nucleus (or areola as he called it) of the cell, was not confined to the epidermis, being also found, in the pubescence of the surface and in the parenchyma or internal cells of the tissue. This nucleus of the cell was not confined to only orchids, but was equally manifest in many other monocotyledonous families and in the epidermis of dicotyledonous plants, and even in the early stages of development of the pollen.

In some plants, as Tradascantia virginica, it was uncommonly distinct, especially in the tissue of the stigma, in the cells of the ovum, even before impregnation, and in all the stages of formation of the grains of pollen.

For the first 150 years, the cell theory was primarily a structural idea. This structural view, which is found in most textbooks, describes the components of a cell and their fate in cell reproduction. Since the 1950's, however, cell biology has focused on DNA and its informational features. Today we look at the cell as a unit of self-control, i.e.., the description of a cell must includes ideas about how genetic information is converted to structure.

Structure of Prokaryotic and Eukaryotic Cells

Living cells are divided into two types - prokaryotic and eukaryotic (sometimes spelled procaryotic and eucaryotic). This division is based on internal complexity.

Eukaryotic Cells: The cells of protozoa, higher plants and animals are highly structured. These cells tend to be larger than the cells of bacteria, and have developed specialised packaging and transport mechanisms that may be necessary to support their larger size. Here are representive drawings of plant and animal cells showing their respective organelles.

Eukaryotes (also spelled "eucaryotes") are organisms with complex cells, in which the genetic material is organised into membrane-bound nuclei.

They include the animals, plants, and fungi, which are mostly multicellular, as well as various other groups collectively referred to as protists, many of which are unicellular. In contrast, other organisms such as bacteria lack nuclei and other complex cell structures, and are called prokaryotes. The eukaryotes share a common origin, and are often treated formally as a super-kingdom, empire, or domain. The name comes from the Greek eus or true and karyon or nut, referring to the nucleus.

Structure: Eukaryotic cells are generally much larger than prokaryotes, typically a thousand times by volume. They have a variety of internal membranes and structures, called organelles, and a cytoskeleton composed of microtubules and micro-filaments, which play an important role in defining the cell's organisation. Eukaryotic DNA is divided into several bundles called chromosomes, which are separated by a micro-tubular spindle during nuclear division. In addition to asexual cell division, most eukaryotes have some process of sexual reproduction via cell fusion, which is not found among prokaryotes.

Animal cells are typical of the eukaryotic cell, enclosed by a plasma membrane and containing a membrane-bound nucleus and organelles. Unlike the eukaryotic cells of plants and fungi, animal cells do not have a cell wall. This feature was lost in the distant past by the single-celled organisms that gave rise to the kingdom Animalia. Most cells, both animal and plant, range in size between 1 and 100 micrometres and are thus visible only with the aid of a microscope.

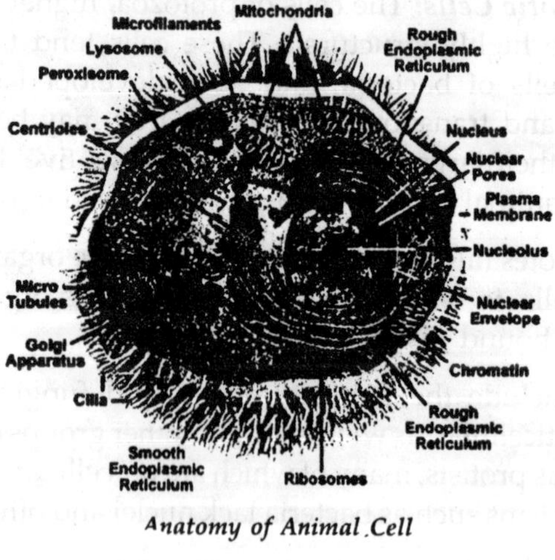

Anatomy of Animal Cell

Membranes: Eukaryotic cells include a variety of membrane-bound structures, collectively referred to as the endomembrane system. Simple compartments, called vesicles or vacuoles, can form by budding off other membranes. Many cells ingest food and other materials through a process of endocytosis, where the outer membrane invaginates and then pinches off to form a vesicle. It is probable that most other membrane-bound organelles are ultimately derived from such vesicles. The nucleus is surrounded by a double membrane, with pores that allow material to move in and out.

Various tube- and sheet-like extensions of the nuclear membrane form what is called the endoplasmic reticulum or ER, which is involved in protein transport. It includes rough sections where ribosomes are attached, and the proteins they synthesise enter the interior space or lumen. Subsequently, they generally enter vesicles, which bud off from the smooth section. In most eukaryotes, the proteins may be further modified in stacks of flattened vesicles, called Golgi bodies or dictyosomes.

Vesicles may be specialised for various purposes. For instance, lysosomes contain enzymes that break down the contents of food vacuoles, and peroxisomes are used to break down peroxide which is toxic otherwise. Many protozoa have contractile vacuoles, which collect and expel excess water, and extrusomes, which expel material used to deflect predators or capture prey. In multicellular organisms, hormones are often produced in vesicles. In higher plants, most of a cell's volume is taken up by a central vacuole or tonoplast, which maintains its osmotic pressure.

Mitochondria and plastids: Mitochondria are organelles found in nearly all eukaryotes. They are surrounded by double membranes, the inner of which is folded into invaginations called cristae, where aerobic respiration takes place. They contain their own DNA and are only formed by the fission of other mitochondria. They are now generally held to have developed from endosymbiotic prokaryotes, probably proteobacteria. The few protozoa that lack mitochondria have been found to contain mitochondrion-derived organelles, such as hydrogenosomes and mitosomes.

Plants and various groups of algae also have plastids. Again, these have their own DNA and developed from endosymbiotes, in this case cyanobacteria. They usually take the form of chloroplasts, which like cyanobacteria contain chlorophyll and produce energy through photosynthesis. Others are involved in storing food. Although plastids likely had a single origin, not all plastid-containing groups are closely related. Instead, some eukaryotes have obtained them from others through secondary endosymbiosis or ingestion. Endosymbiotic origins have also been proposed for the nucleus and eukaryotic flagella, but this is not generally accepted, both from a lack of cytological evidence and difficulty in reconciling this with cellular reproduction.

Cytoskeletal Structures: Many eukaryotes have slender motile projections, usually called flagella when long and cilia when short, that are variously involved in movement, feeding, and sensation. These are entirely distinct from prokaryotic flagella. They are supported by a bundle of microtubules arising from a basal body, also called a kinetosome or centriole, characteristically arranged as nine doublets surrounding two singlets. Flagella also may have hairs or mastigonemes, scales, connecting membranes and internal rods. Their interior is continuous with the cell's cytoplasm.

Centrioles are often present even in cells and groups that do not have flagella. They generally occur in groups of one or two, called kinetids, that give rise to various microtubular roots. These form a primary component of the cytoskeletal structure, and are often assembled over the course of several cell divisions, with one flagellum retained from the parent and the other derived from it. Centrioles may also be associated in the formation of a spindle during nuclear division. Some protists have various other microtubule-supported organelles.

These include the radiolaria and heliozoa, which produce axopodia used in flotation or to capture prey, and the haptophytes, which have a peculiar flagellum-like organelle called the haptonema.

Reproduction: Nuclear division is often coordinated with cell division. This generally takes place by mitosis, a process which allows each daughter nucleus to receive one copy of each chromosome. In most eukaryotes there is also a process of sexual reproduction, typically involving an alternation between haploid generations, where only one copy of each chromosome is present, and diploid generations, where two are present, occurring through nuclear fusion (syngamy) and meiosis. There is considerable variation in this pattern, however.

Origin and evolution: The origin of the eukaryotic cell was a milestone in the evolution of life. Although eukaryotes use the same genetic code and metabolic processes as prokaryotes, their higher level of organisational complexity has permitted the development of truly multicellular organisms. Without eukaryotes, the world would lack mammals, birds, fish, invertebrates, mushrooms, plants, and complex single-celled organisms. Eukaryotes are generally held to have originated around 2 billion years ago, but this has been disputed. Early fossils such as acritarchs are difficult to interpret. Forms that can be related to modern groups start appearing around 800 million years ago, and most fossil lines are known by the end of the Cambrian, around 500 million years ago.

Genetic studies during the 1980s and 1990s left most eukaryotes in an unresolved "crown" group, usually divided by the form of the cristae in their mitochondria. The few amitochondriate groups branched first on rRNA trees and so were considered basal, but this is now considered to be an error caused by long branch attraction. A new picture has been slow to develop. Most eukaryotes are now included in several supergroups:

Opisthokonts	Animals, fungi, choanoflagellates, etc.
Amoebozoa	Most lobose amoebae and slime moulds
Rhizaria	Various amoeboid protozoa
Excavates	Various flagellate protozoa
Plantae sensu lato	Land plants, green and red algae
Chromista	Brown algae, diatoms, water moulds, etc.
Alveolates	Ciliates, Apicomplexa, dinoflagellates, etc.

However, some protists are not closely related to any of these lines, and the relationships between the different supergroups remain almost entirely uncertain. In particular, there is dispute about where the root of the evolutionary tree belongs, and as a result what the earliest eukaryotes were like.

Prokaryotic Cells: These cells are simple in structure, with no recognisable organelles. They have an outer cell wall that gives them shape. Just under the rigid cell wall is the more fluid cell membrane. The cytoplasm enclosed within the cell membrane does not exhibit much structure when viewed by electron microscopy.

Prokaryotes are organisms without a cell nucleus (= karyon), or indeed any other membrane-bound organelles, in most cases unicellular (in rare cases, multicellular). This is in contrast to eukaryotes (also spelled "eucaryotes"), organisms that have cell nuclei and may be variously unicellular or multicellular. The difference between the structure of prokaryotes and eukaryotes is so great that it is considered to be the most important distinction among groups of organisms. Most prokaryotes are bacteria, and the two terms are often treated as synonyms. However, Woese has proposed dividing prokaryotes into the Bacteria and Archaea (originally Eubacteria and Archaebacteria) because of the significant genetic differences between the two. This controversial arrangement of Eukaryote, Bacteria, and Archaea is called the three-domain system.

Structure: The cell structure of prokaryotes differs greatly from eukaryotes in many ways. The defining characteristic is, of course, the absence of a nucleus or nuclear envelope. Prokaryotes also were previously considered to lack cytoskeletons and do lack membrane-bound cell compartments such as vacuoles, endoplasmic reticulum, mitochondria and chloroplasts. In eukaryotes, the latter two perform various metabolic processes and are believed to have been derived from endosymbiotic bacteria. In prokaryotes similar processes occur across the cell membrane; endosymbionts are extremely rare.

Prokaryotes also have cell walls, while some eukaryotes, particularly animals, do not. Both eukaryotes and prokaryotes have structures called ribosomes, which produce protein. Prokaryotes are usually much smaller than eukaryotic cells. Prokaryotes have a single circular (only exceptionally linear, as in Borrelia burgdorferi) chromosome, contained within a region called nucleoid, rather than in a membrane-bound nucleus, but may also have various small circular pieces of DNA called plasmids spread throughout the cell. Reproduction is exclusively asexual, through binary fission, where the chromosome is duplicated and attaches to the cell membrane, and then the cell divides in two.

However, they show a variety of parasexual processes where DNA is transferred between cells, such as transformation and transduction. While prokaryotes are nearly always unicellular, some are capable of forming groups of cells called colonies. Unlike many eukaryotic multicellular organisms, each member of the colony is undifferentiated and capable of free-living.

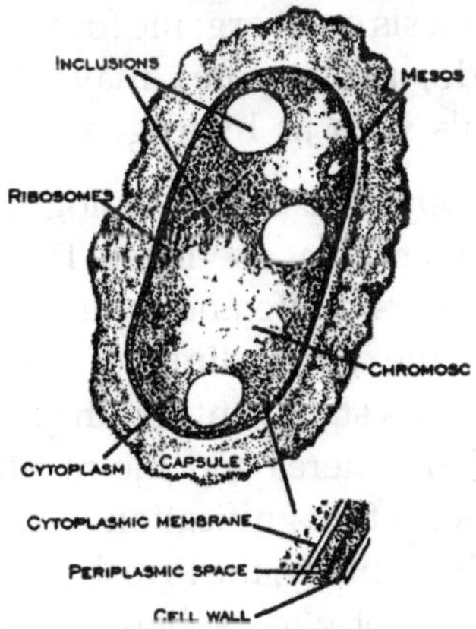

Structure of prokaryotic cells

Colonies are formed by organisms that remain attached following cell division, sometimes through the help of a secreted slime layer. At least some prokaryotes contain protein-enclosed micro-compartments which can be seen as primitive organelles.

Environment: Prokaryotes are found in nearly all environments on earth. Archaea in particular seem to thrive in harsh conditions, such as high temperatures or salinity. Organisms such as these are referred to as extremophiles. Many prokaryotes live in or on the bodies of other organisms, including humans. Sometimes this leads to a life-threatening bacterial infection, but in many cases the organisms are harmless or even beneficial to the host.

Evolution of prokaryotes: It is generally accepted that the first living cells were some form of prokaryote. Fossilised prokaryotes 3.5 billion years old have been discovered, and prokaryotes are perhaps the most successful and abundant organism even today. While earth is the only known place where prokaryotes exist, some have suggested structures within a Martian meteorite should be interpreted as fossil prokaryotes, but this is extremely doubtful. Prokaryotes diversified greatly throughout their long existence. The metabolism of prokaryotes is far more varied than that of eukaryotes, leading to many highly distinct types of prokaryotes. For example, in addition to using photosynthesis or an organic form of carbon for energy like eukaryotes do, prokaryotes may obtain energy from inorganic chemicals such as hydrogen sulphide.

Light Microscopy: For many biologists, the microscope belongs to the gadgets they use most. The reason is obvious: cells are the basic units of life and their size is, with only few exceptions, below the upper limit of the resolution of the human eye. Numerous statements on the refraction of light by lenses or waterfilled spheres and the resulting magnification proof, that the effects of magnification were already known in antiquity. In the 13th century, the British natural scientist Roger Bacon used segments of glass spheres as magnifying glasses and recommended them to partially sighted persons as spectacles. The grinding of lenses was perfected in the following centuries.

The term microscope was introduced by members of the Italian Accademia de Lincei (Academy of The Lynx), whose most prominent member was G. Galilei. Independent of the beginning development of the two-lensed - also called the compound microscope - stronger and stronger lenses became fashionable, too.

In 1667, the "Micrographia" of the British natural scientist Robert Hooke was published. Hooke examined vegetable tissues with the help of a compound microscope and thus detected their basic elements, the cells. The further development of microscopes and especially the use of microscopes in the 18th century went on rather hesitantly. This was on one hand caused by the opinion of many natural scientists of reputation, that a lot had still to be discovered even without the use of microscopes. On the other hand, the interfering chromatic aberrations were a continuing source of misinterpretations. This led to the prejudice, that anything you wanted to see could be seen within a microscope.

As soon as 1695, D. Gregory considered the combination of several lenses with differing dispersions in order to minimise this handicap. The mathematician L. Euler brought forth a theoretical explanation of achromates (systems of lenses with corrected colours) in 1771 and he suggested the calculation of such achromatic objectives. They were finally built by F. G. Beeldsnyder, who placed a biconcave lens out of flint glass between two biconvex lenses out of another type of glass, that had a different index of refraction.

G. D. Amici developed achromatic systems of lenses further and invented also immersion objectives. The actual break-through on the way to the construction of modern microscopes was due to the formulation of a theory of image formation within the microscope by E. ABBE. He proved, that an absolute upper limit of microscopic image formation exists and that it was dependent of the aperture of the objective and the wave length of the light. His theory permitted the development of reproducible high-performance microscopes and their production in series.

The use of newly developed types of glasses by O. Schott led in 1886 to the development of apochromatic objectives (objectives, that remove a rest of colours, the so-called secondary spectrum) by E. ABBE and C. Zeiss. The progress of microscope building enabled a quick series of numerous discoveries in the subjects of histology, cytology and bacteriology. The progress was helped by the development and use of suitable methods of fixation, embedding and cutting, specific dyes and conservatives.

Dark Field Microscopy: A main obstacle in the microscopy of biological objects is their poor contrast. Only where a contrast exists or where it can be achieved by contrast-enhancing dyes, structures can be made visible. Light-absorbing parts of a preparation weaken the amplitude of the light waves that pass through them. It is thus also spoken of amplitude preparations. The change of stronger and feebler light is perceived by the eye as a difference in brightness. The invisible parts of the preparation are went through by the light without a change of amplitude, but the phase of the light may be changed depending on the consistency of the material. This change is due to the altered speed of the lightwaves. Differences in phase can be perceived neither by the eye nor by a photographical film.

The Dutch physicist F. ZERNIKE succeeded in 1935 to convert phase to amplitude differences. He was awarded the Nobel prize for this achievement in 1953. Today his method is known as phase-contrast microscopy and it is by now an integral part of nearly all research and many teaching microscopes. Its eminent advantage is that it allows the examination of living objects and thus to follow the processes within cells. It was the phase-contrast microscope that made it possible to make mitosis visible and even to film it.

For the procedure itself a special condenser with a ring-shaped mask and an additional "phase-ring" that is fixed within the back focal plane of the objective is needed. The "phase-ring" has two important tasks:

1. It has to achieve an alignment of the brightness of refracted and unrefracted light as the rays that pass through the preparation are weakened in their intensity. In contrast to a conventional image gained with the help of a light microscope, the background of a phase-contrast image is thus dark.
2. The phase difference of most biological preparations is one-quarter of a wavelength or less (lambda /4). The phase-ring is constructed so as to achieve an additional difference of lambda / 4. The complete phase difference is thus lambda / 2 and crest and trough of refracted and unrefracted rays extinguish each other. A disadvantage of this method is the appearance of light halos around some objects ("halo-effect").

Fluorescence Microscopy: Fluorescence microscopy is based on the fact that some molecules emit part of the light absorbed by them as longer waves. A well-known example is the red fluorescence of chlorophyll. Since the Austrian teacher M. HAITINGER made his examinations in the thirties of this 20th century it is known that a number of so-called fluorochromes exist, with which microscopic preparations can be stained so that they emit fluorescence indirectly. In the following decades, a whole range of so-called vital dyes were detected or developed, that -used in low concentrations- could mark specific parts of living cells or of tissues. They enabled researchers to follow the solute transport within cells and tissues or to ascertain the pH of special compartments, like for example the vacuole.

Since roughly twenty years fluorescence microscopy is flowering again, on one hand, because the range of fluorochromes became much broader and on the other hand, because completely new approaches like indirect fluorescence could be used successfully. Additionally the construction of microscopes was improved and highly efficient masks were developed. There are two ways in which fluorescence microscopes can be constructed: as epifluorescence- and as transmission microscopes. The second is the older way of construction. Three components are needed:

1. A strong source of light, that emits mainly short lightwaves. Mercury high pressure lights have proven useful.
2. The first barrier filter: This filter helps to shut off all radiation other than the one that activates the specific dye. It is placed behind the light source within the light cone. Besides it is advantageous to work with a dark field condenser.
3. Second barrier filter: This filter is brought into the light cone between objective and eyepiece. It lets through only long wavelengths, that are caused by emission of the preparation (so-called "secondary radiation").

In the last years the epifluorescence microscope has replaced the transmission fluorescence microscope more and more. But the transmission fluorescence microscope is still better suited to only weakly magnifying objectives (2.5 x, 6.3 x). With the epifluorescence the objective is also a condenser and the stronger it is the more intensive radiation can be used. The heart of epifluorescence is a construction within the light cone between objective and eyepiece that serves to feed activating radiation into the system and is constructed of first barrier filter, beam-splitting mirror and second barrier filter.

Polarising and Interference Contrast Microscopy: Light waves do usually radiate into all directions. Suitable polarisation masks allow the elimination of certain planes of polarisation, whereby linear polarised light is gained. It can be completely erased by a second polarisation mask that is placed at a right angle to the first one. Such polarisation masks that can rotate around their own axis can be placed in the light cone of a microscope. One beneath the condenser and the other above the objective.

The use of a thus constructed polarisation microscope is only sensible, if preparations with polarising properties are examined. This is the case, if the preparation consists of units that are oriented in a certain way like molecules or atoms (double refracting crystals). This type of microscope is thus used mainly in mineralogy since crystals are per definitions built in a regular pattern. By the use of polarised light crystal axis and three-dimensional can pattern be ascertained exactly. The practical application in biology is somewhat limited, but the structure of starch grains for example or the orientation of cellulose fibrils within the vegetable cell wall or the orientation of rod-shaped viruses (tobacco mosaic virus) can be determined.

Based on polarisation microscopy the French physicist G. Nomarsky developed the interference contrast microscope (also called differential interference contrast, DIC) in Paris during the middle of the 1950s. For this type of microscopy two additional Wollaston-prisms are needed. A Wollaston-prism is composed of two cemented calcium fluoride wedges. A polarised ray of light is split into two beams at the cemented plane that are vertical towards each other. The first Wollaston-prism is used in the front focal plane of the condenser, the second in the back focal length of the objective. The object is thus passed by two beams that are vertical to each other.

The phase of the beams is shifted according to the thickness and the refraction characteristics of the object. Optimal interference contrast is reached at the rims of the object, where the phase of both beams is shifted differently. It is important, however, how the object is orientated. A rotating microscope stage is used to analyse the object in all its orientations. The second Wollaston-prism reunites the rays. To achieve interference the planes of oscillation have to coincide. This is done, by the polarisation mask above the objective.

An interference contrast image looks like a three-dimensional relief. An inexperienced microscoper could thereby be lead to the misapprehension of a three-dimensional structure of the preparation. This is not the case. The tree-dimensional appearance stems from the fact that the differences in the density of the preparation are transformed into differences in height in the image. In contrast to phase-contrast microscopy even relatively thick preparations can be analysed. Besides the light-dark-contrasts, that can be enhanced or diminished by the rotation of the polarisator or by the adjustment of the second Wollaston-prism, colour contrasts can be achieved by the use of a lambda / 4 mask.

Ectron Microscopy: In 1924 L. de Broglie discovered the wave-character of electron rays thus giving the prerequisite for the construction of the electron microscope. The prototype was built by M. Knoll and E. Ruska. One of the first biological objects depicted was the tobacco mosaic virus (TMV).

Transmission Electron Microscopy (TEM): The conventional electron microscopy is nowadays called TEM (transmission electron microscopy). We will therefore start with its construction. The ray of electrons is produced by a pin-shaped cathode heated up by current. The electrons are vacuumed up by a high voltage at the anode. The acceleration voltage is between 50 and 150 kV.

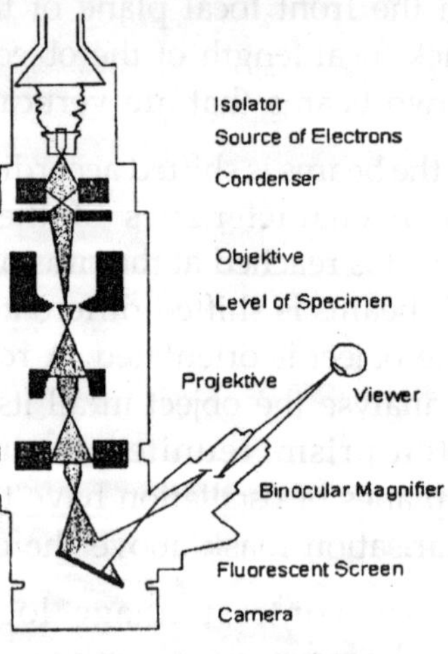

Structure of a Microscope

The higher it is, the shorter are the electron waves and the higher is the power of resolution. But this factor is hardly ever limiting. The power of resolution of electron microscopy is usually restrained by the quality of the lens-systems and especially by the technique with which the preparation has been achieved.

Modern gadgets have powers of resolution that range from 0,2 - 0,3 nm. The useful resolution is therefore around 300,000 x. The accelerated ray of electrons passes a drill-hole at the bottom of the anode. Its following way is analogous to that of a ray of light in a light microscope. The lens-systems consist of electronic coils generating an electromagnetic field. The ray is first focused by a condenser. It then passes through the object, where it is partially deflected. The degree of deflection depends on the electron density of the object.

The greater the mass of the atoms, the greater is the degree of deflection. Biological objects have only weak contrasts since they consist mainly of atoms with low atomic numbers (C, H, N, O). Consequently it is necessary to treat the preparations with special contrast enhancing chemicals (heavy metals) to get at least some contrast. Additionally they are not to be thicker than 100 nm, because the temperature is raising due to electron absorption. This again can lead to destruction of the preparation. It is generally impossible to examine living objects.

After passing the object the scattered electrons are collected by an objective. Thereby an image is formed, that is subsequently enlarged by an additional lens-system (called projective with electron microscopes). The thus formed image is made visible on a fluorescent screen or it is documented on photographic material. Photos taken with electron microscopes are always black and white. The degree of darkness corresponds to the electron density (= differences in atom masses) of the candled preparation.

Scanning Electron Microscope (SEM): The path of the electron beam within the scanning electron microscope differs from that of the TEM. The technology used is based on television techniques. The method is suitable for the depiction of preparations with conductive surfaces. Biological objects have thus to be made conductive by coating with a thin layer of heavy metal (usually gold is taken). The power of resolution is normally smaller than in transmission electron microscopes, but the depth of focus is several orders of magnitude greater. Scanning electron microscopy is therefore also well-suited for very low magnifications.

The surface of the object is scanned with the electron beam point by point whereby secondary electrons are set free. The intensity of this secondary radiation is dependent on the angle of inclination of the object's surface. The secondary electrons are collected by a detector that sits at an angle at the side above the object. The signal is then enhanced electronically. The magnification can be chosen smoothly and the image appears a little later on a viewing screen.

Preparation Techniques: Biological objects can only be examined with the electron microscope after long and careful preparations. Years of research and results achieved with different methods finally guaranteed the reliability of the methods and results. Biological preparations - even single cells- are usually too big and too thick to be used as a whole. Normally cross-sections have to be prepared. Sections require the following steps:

1. Fixation of the material, usually with glutaraldehyde (covalent cross-linking of protein molecules) and osmium tetroxide (binds to and stabilises membranes). Dehydration of the specimen.
2. Permeation with a monomeric resin that polymerises to form a solid block of plastic. Without the thus strengthened structures the specimen would collapse in the vacuum of the microscope.
3. Cutting of the specimen: needed is an ultramicrotome that can produce sections of about 15-100 nm thickness

(about 1/200 of the thickness of a cell). Ultra-microtomes do normally have fine glass knifes. Edges of break of glass are sharper than metal knifes, but they do not last very long. Diamond knifes are an alternative. They live longer, but are also much more expensive.

4. The thin sections are placed on a small circular metal grid that is coated with a coal-strengthened plastic (formvar) for viewing in the microscope.
5. Two different methods for contrast enhancement exist: coating and impregnation with heavy metal ions. Coating is achieved by placing the specimen in a vacuum and exposing it to a cloud of metal dust (platinium, platinium/coal, gold, vanadium, chrome, lead, etc.). The cloud is produced by the heating of a metal filament that is placed at a defined distance and angle from the specimen. The relief-like surface of the specimen helps the formation of irregular structured metal coats on the specimen (relief contrast). The imprint of a specimen can also be taken. Therefore slightly thicker metal coats are prepared that are taken off the specimen before viewing them in the microscope.

When contrast is achieved with heavy metal ions, then the preparation is impregnated with uranyl acetate- or lead citrate solutions. The salts are absorbed by the specimen with different strength, so that differently labelled structures are viewed in the microscope later on. It is spoken of positive staining, if a special structure has absorbed the ions and of negative staining, if the metal ions (phosphoric tungsten acid, uranyl acetate, uranyl formiate and others) accumulate around the actual structure. Negative staining is normally used to make macromolecules and molecular complexes (ribosomes, viruses) visible. Usually a special chemical has to be added that prevents the molecules from getting lumpy.

Freezing Methods: Freeze methods (like freeze-drying) offer an alternative to chemical fixation with often better preservation of the specimen's structures. Freeze-fracture and freeze-etching are well-suited for small specimen: cells and subcellular structures. The specimen is frozen and the frozen material is "broken" with the aid of a glass knife. The preparation splinters and the edges of breaking run along the membrane or between two half membranes. Water is subsequently removed via freeze drying. It sublimates, i.e. it transforms out of its solid phase directly into the gas phase. This process etches the surface of the preparation. It is then coated and the metal coat is taken off and viewed in the microscope.

During the late 1970s, J. E. Heuser developed the quick-freeze, deep-etch preparation of samples that avoids the crystallisation of water. The (very small) specimens are frozen quickly in liquid nitrogen transferring the water into a glass-like state. This leads to a much better preservation of the sample. The method is suitable for the depiction of large molecules and molecular complexes. Here, too, an imprint is gained by coating. Contrary to negative staining a three-dimensional image of the specimen can be formed.

10

Cell Walls (Plants)

Plant cells are separated from each other by a substantial extracellular matrix, the cell wall. Although the cell wall is not impervious, it is likely to hinder cell-to-cell communication.

Moreover, because plant cells cannot migrate during development, location, instead of lineage, is the primary determinant of cell fate in plants, making communication over both long and short distances essential for the coordination of plant growth. It appears that plants have significantly overcome the downside of having a cell wall by forming channels between cells to allow the transit of signalling and other important molecules. Plants may even be considered as supracellular organisms, in that whole tissues are symplastically connected.

The channels that connect plant cells are called plasmodesmata (PD), and their investigation offers tantalising clues as to how plant cells may use them to communicate with each other and to coordinate development. Although once seen as constricted channels through which only molecules <1 kD in size could pass, PD have recently been shown to be far more dynamic and allow passage of both proteins and nucleic acids. Other findings, such as the observation that transcription factors can move between cells, and that RNA movement may be behind a range of long-distance signals in plants, raise further questions regarding a regulatory role of PD in development.

Plasmodesmata

Ultrastructure: Ultrastructural studies of the PD by transmission electron microscopy (TEM) have shown them to be membrane-membrane-lined pores, with an exterior membrane contributed by the plasma membranes of the two cells connected by the PD, and a central desmotubule, derived from endoplasmic reticulum that is also contiguous between the two cells. Typically, PD have a small diameter of –40 nm, with the central desmotubule being connected to the plasma membrane via spoke-like extensions, resulting in microchannels having an estimated diameter of –2.5 nm.

It is important to bear in mind that these measurements are all based on fixed tissue, and that nothing is known from these observations about the dynamic behaviour of PD proteins. Morphologically, two distinct forms of PD have been recognised. Primary PD are formed at the cell plate, a plant-typical structure that consists of the incipient plasma membrane and cell wall synthesised during cytokinesis. Secondary PD, in contrast, are formed between existing cell walls. Both forms of PD can be either single-channels or branched. It is not known whether these different forms of PD have distinct functions.

Molecules are thought to be transported through the transport channels (TC) between the appressed ER and plasma membrane (PM). Spoke-like structures (S) connect the appressed ER to the PM. Below is a transverse cross-section of the same structure.

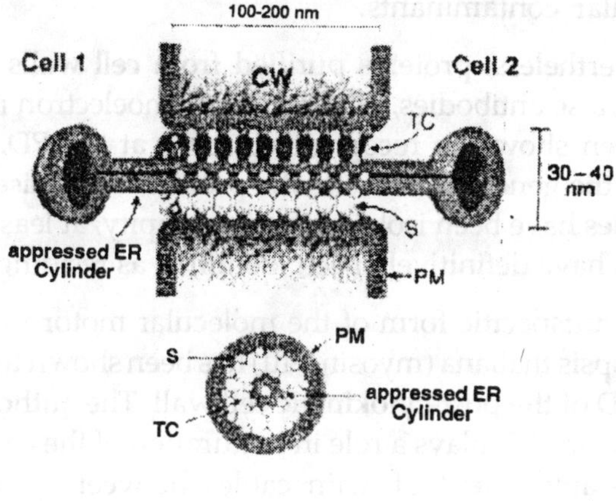

A longitudinal schematic cross-section through a PD.

Proteins of the PD: A complete understanding of how the PD function to control movement of molecules between plant cells is dependent on identifying the proteins of which they are composed. Genetic screens to identify mutants defective in intercellular movement are hindered by the fact that PD components are probably essential for viability. Because it is questionable whether one can recover viable mutants that are seriously defective in the structure or function of PD, it will probably be essential to study mutants at the embryo stage of development.

One such screen is briefly mentioned in a recent review. It has led to the identification of an embryonic lethal mutant, ise-1, in which tracer molecules of up to 11 kD can easily move between cells, which is a substantial increase over the wild-type size exclusion limit of <3kD. The disrupted gene is a promising candidate for a protein that regulates the aperture of PD. A complementary approach to the genetic dissection of PD function is to biochemically purify these organelles from cell walls and then to identify the PD constituents directly. Although potentially powerful, this approach has so far been hampered by the small amount of protein present and the difficulty in extracting intact PD from the cell wall without subcellular contaminants.

Nevertheless, proteins purified from cell walls have been used to raise antibodies, which by immunoelectron microscopy have been shown to recognise epitopes at the PD. However, none of the genes that encode the proteins recognised by these antibodies have been isolated. By serendipity, at least two plant proteins have definitively been identified as residing at the PD.

A plantspecific form of the molecular motor myosin from Arabidopsis thaliana (myosin VIII) has been shown to be present in the PD of the post-cytokinetic cell wall. The authors propose that myosin VIII plays a role in maturation of the cell plate and the reestablishment of actin cables between cells to allow intercellular communication. The same group has also found by immunoelectron microscopy that the multifunctional calcium-sequestering protein calreticulin is enriched in ER domains associated with the PD of maize.

Because myosin function is regulated by calcium binding, the authors have speculated that calreticulin regulates myosin VIII and thereby provides a Ca^{2+}-dependent gating mechanism for PD. Other cytoskeletal antigens have been identified as potential PD components by microscopy as well, including actin and myosin in the alga Chara corallina, as well as actin-like and myosin-like molecules in diverse plant species.

Size Exclusion Limits: The narrow aperture of the PD seen in the TEM suggested that PD would only allow the passage of very small molecules. Injection studies with fluorescently labelled dextrans generally supported the view of the PD as static channels through which only molecules of <1 kD could diffuse. Additionally, there appear to be boundaries that limit the diffusion even of small dyes, thereby dividing tissues into so-called symplasmic domains.

Interestingly, the extent of symplasmic domains as determined by injection of fluorescent dyes is under developmental regulation, even though the biological significance is unknown. The dogma that PD allow only the diffusion of small molecules has been challenged from several sides during the past few years, with data showing that the PD are dynamic in nature and that they allow the non-specific trafficking of numerous proteins.

Researchers used a non-invasive transfection system to show that a red-shifted version of GFP (27 kD) moved readily in tobacco plant leaves, and they even found limited movement of 2xGFP (54 kD). In these and other studies, the size exclusion limit (SEL) has been shown to be closely related to the developmental state of the tissue. When tobacco leaves mature and change from being net consumers of carbon (sink tissue) to being carbon exporters (source tissue), there is a dramatic decrease in the SEL.

Whereas molecules of up to 50 kD may move freely in sink tissue, movement is greatly reduced in the source tissue. This change in SEL is accompanied by a change of PD morphology from simple to branched. In a complementary study, GFP was expressed under a promoter specific for so-called companion cells, which line the phloem, the part of the vasculature that transports assimilates. In these transgenic plants, GFP could move over long distances through the phloem into leaves and other sink tissues.

Viral Movement Proteins, and the Plant Cytoskeleton: A major contribution to our understanding of PD function comes from the investigation of plant viruses. For viruses to successfully infect plants, they must be able to spread from cell to cell, a process that is greatly restricted by the presence of cell walls. To bypass these formidable barriers, viruses have evolved highly efficient mechanisms that exploit the cytoskeleton within plant cells, and that allow them to traffic in a rapid targeted fashion throughout the plant.

Many plant viruses encode movement proteins that actively target the PD and that are essential for viral infectivity. In the absence of these proteins, viruses will replicate and encapsidate their genomes, but they are unable to spread from cell to cell. Movement proteins have many interesting properties, and they have become essential tools for studying intercellular transport. Plants contain endogenous proteins that are antigenically related to viral movement proteins and that may share similar functions.

The best studied viral protein is the 30-kD movement protein from tobacco mosaic virus (TMV MP30). Apart from overexpression studies, which first showed that MP30 can increase the SEL of PD, the analysis of GFP fusions by light microscopy has revealed a large amount of information about movement protein function. The tagged proteins are seen to interact with the ER, the actin and microtubule cytoskeleton, as well as with the PD themselves.

Consistent with the apparent microtubule association, MP30 has been found to have a conserved motif for microtubule binding. It has been proposed that the binding of MP30 to the sides of microtubules enables it to be actively transported between cells. Consistent with this hypothesis, disruption of the interaction between MP30 and the cytoskeleton prevents both viral movement and infectivity. The association of MP30 with the ER is likely related to the fact that viral replication occurs on the ER. It is thought that, in some cases, the viral movement protein associates with viral RNA, and that this complex is transported along the cytoskeleton to the PD, through which the virus is delivered to adjacent cells.

Other studies indicate that movement proteins not only are able to target PD, but also actively dilate the PD, enabling the increased movement of other proteins in infected cells. It has thus been proposed that PD may be considered as having three states: closed, open, and dilated. Based on the argument that viral movement through PD must require plant host factors, several groups have taken biochemical and reverse genetic approaches to dissect viral movement. These efforts indicate the involvement of at least two cell wall-associated enzymes in viral movement and thus presumably PD function.

Two independent screens led to the identification of pectin methylesterase, an enzyme that modifies the pectin in cell walls, as a protein that directly interacts with TMV MP30 in vitro. In another study, it was found that reduction of β-1,3-glucanase activity in transgenic plants both decreased the SEL and slowed intercellular viral movement. The mechanisms by which these host factors may affect PD function need to be further clarified.

RNA and Proteins: Movement in Plants

Most of the early studies on PD function were carried out using microinjection or biolistic bombardment to deliver tracer molecules, and the results from these reports have been recently reviewed in detail. Here focus on movement observed using genetic chimeras, including grafts and periclinal chimeras, and tissue-specific transgenic expression of signalling molecules.

In the figure below (A) In grafting experiments, a scion is grafted onto a stock that is genetically different, for example, because it carries a transgene. If the phenotype is graft-transmissible, the scion will develop the same phenotype as the stock owing to movement of the signal through the phloem.

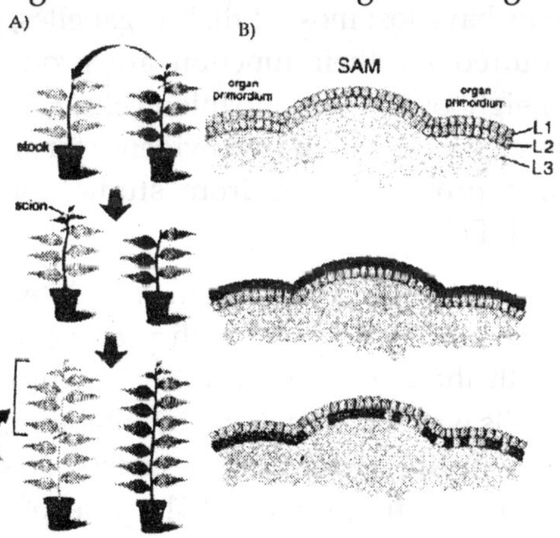

Diagrams of different methods to generate genetic chimeras.

(B, top) Cross-section of a shoot apex, showing the three tissue layers. The shoot apical meristem (SAM) is at the centre, with emerging organ primordia on both sides. (Middle) Localised ectopic expression can be driven by an epidermis-specific promoter in L1 only. (Bottom) Induced genetic recombination can result in clones of marked cells in L2 with different genetic composition.

Long-distance Movement along the Phloem: Long-distance transport of a variety of molecules, including but not limited to water, nutrients, hormones, and other signals, is essential for plant growth. Long-distance transport relies on the plant vascular system, which includes two tissues, the xylem and the phloem. The xylem mainly consists of files of dead cells, and the flow within it is driven by a tension gradient. On the other hand, the sieve elements of the phloem are formed by files of living, but enucleate, cells.

The sieve elements and the surrounding companion cells descend from the same mother cells, and they are connected by a large number of plasmodesmata, which have a larger SEL compared to the ones found in mature leaves. Because the sieve elements have lost most of their organelles, proteins and mRNAs required for their function are produced in their immediate neighbours, the companion cells, then transported into the sieve elements. The clearest evidence for an involvement of PD in this process comes from studies of the sucrose transporter SUT1.

Both SUT1 mRNA and protein are detected in sieve elements by in situ localisation, with preferential association of mRNA with the PD. That the mRNA is produced in the companion cells was shown directly by inhibiting its synthesis in these cells using an antisense construct driven by a companion-cell-specific promoter. The role of PD between companion cells and sieve elements in gating the up-loading and unloading of these and other molecules into the phloem has been discussed in detail by Ruiz-Medrano et al.. In the past few years, long-distance movement of RNA has caught much attention.

Using grafting experiments with transgenic and wild-type plants, Palauqui and Vaucheret were the first to show that post-transcriptional silencing, a syndrome that includes RNA interference (RNAi), can spread to cells that do not contain the causal transgene. The signal consists most likely of small interfering RNAs that are transported through the phloem, and leads to the silencing of genes bearing sequence homology to the RNA molecules. This mechanism provides an important line of defence against a number of viruses, many of which have in turn developed strategies to render this mechanism ineffective.

Apart from these small RNAs, several mRNAs have been isolated from phloem sap of pumpkin and shown by grafting experiments to traffic for long distances via the phloem. Some of these RNAs encode proteins that are related to viral movement proteins, but others encode proteins with similarity to developmental regulators. In another elegant grafting experiment, the effects of longdistance mRNA movement have recently been shown. A naturally occurring mutation in tomato, Mouse ear (Me), is the result of a gene fusion in which regulatory sequences of the PYRO-PHOSPHATE-DEPENDENT PHOSPHO-FRUCTOKINASE (PFP) gene cause high-level expression of the homeobox gene LeT6, resulting in altered leaf morphology.

Researchers showed that the Me mRNA can move from mutant stocks into wild-type scions, with the consequence that genetically wild-type scions show the Me phenotype. This observation strongly supports the assertion that long-distance movement of mRNAs is used as a developmental signal.

Short-distance Movement of Transcription Factors: Plant growth differs from that of animals in that most organs originate post-embryonically from meristems, which are groups of undifferentiated cells that are set aside during embryogenesis.

The aerial parts of the plants are generated by the shoot apical meristem (SAM), which consists of three tissue layers: the outer most layer, L1; the subsurface layer, L2; and the inner layer, L3. These three layers produce the epidermis, subepidermis, and inner tissues such as the vasculature, respectively. In dicotyledonous plants, cells in the L1 and L2 divide almost exclusively in a single plane, with the result that cells in each layer are normally clonally related.

Cells within the same layer are therefore connected mostly via primary PD, and cells from adjacent layers via secondary PD. Because all three layers contribute to cell proliferation and organ differentiation, extensive intercellular communication is required to coordinate their growth. The existence of cross talk between layers has been shown in several plant species using periclinal chimeras, in which neighbouring layers have different genetic composition. Among the genes that have been shown to have non-cell-autonomous functions are several transcription factors controlling flower development.

Initially, it was assumed that their non-cell-autonomous effects were carried out by downstream effectors, such as secreted molecules, because transcription factors were not known to be secreted or move otherwise between adjacent cells. It therefore came as a surprise when Jackson et al. reported that the homeobox-containing transcription factor KNOTTED 1 (KN1) can be detected in the L1 of the maize SAM, even though its RNA is limited to the L2 and L3. The authors insightfully speculated that this discrepancy was caused by KN1 movement, a hypothesis supported by subsequent experiments, in which fluorescently labelled KN1 was injected into tobacco leaf cells.

In this experimental setup, KN1 appeared to generally increase the SEL of PD, and to even effect specific movement of its own sense RNA. Interestingly, the domain required for KN1 movement in tobacco leaves was pinpointed to a potential nuclear localisation signal in the N-terminal part of the homeodomain. Unfortunately, it has not yet been possible to study the effects of KN1 movement within meristem, although KN1 is known to have non-autonomous effects when expressed ectopically. Soon after this first example of transcription factor movement, researchers found that the Antirrhinum MADS domain protein DEFICIENS (DEF) can move within developing flowers. Using genetic chimera, they showed that DEF can move from the L2 into the L1, but not in the other direction.

A third transcription factor shown to move within floral meristems is LEAFY (LFY) in Arabidopsis. The most dramatic results were seen when LFY was expressed from the epidermis-specific ML1 promoter: Although the RNA was, as expected, restricted to the L1, LFY protein was found in a gradient that extended into several interior cell layers. The functionality of trafficked LFY protein was shown both by complete rescue of the mutant phenotype, and by the ability of LFY to activate a direct target gene in interior layers. A caveat of the DEF and LFY studies was that the biological significance of their movement was unclear, because there are no striking differences in their RNA and protein patterns in wild-type plants, although it has been speculated that LFY movement presents a redundant mechanism to ensure that all three tissue layers in the meristem coordinately adopt floral fate.

A more recent publication, however, no longer leaves any doubt as to the significance of transcription factor movement in plant development. Similarly to the shoot, the root is formed by a meristem comprising the root initials. There are distinct initials that give rise to several radially organised layers in the mature root, including from the centre to the periphery, the stele, the endodermis, the cortex, and the epidermis. The SHORT ROOT (SHR) gene, which encodes a member of the GRAS family of transcription factors, is required for the formation of the single-cell-layered endodermis, which descends from the same mother cells as the cortex layer.

Plants lacking SHR activity fail to develop the endodermis layer. In situ hybridisation had revealed that SHR RNA is expressed only in the stele, but subsequent studies using both GFP fusions and immunohistochemistry showed that the protein product is found in both the stele and the adjacent endodermis. Thus, in the wild-type root, SHR protein moves one cell away from its source of expression to the cell layer where its function is required. Using an SHR-GFP fusion protein expressed from its own promoter showed that SHR is both nuclear and cytoplasmic in the stele, but strictly nuclear in the endodermis.

Transcription Factor Movement Mechanisms: Contrary to the common belief that transcription factors should be transported into nuclei as soon as they are made, several transcription factors in plants have been shown to be able to move in developing tissues. How do these proteins go across cellular boundaries? PD seem to be the most suitable answer. The first evidence came from the observation that injected KN1 can change the SEL of tobacco leaf mesophyll cells. An interaction with the PD is further supported by the finding that a KN1 peptide can interfere with KN1's ability to dilate the PD and to move its mRNA into neighbouring cells.

Also, the fact that cells within a layer are typically connected by primary PD but cells in adjacent layers are typically connected by secondary PD suggests a functional basis for the observation that DEF can only move from L2 into L1, but not within a layer. Nevertheless, the mechanisms behind the movement of these proteins remain unknown, and it is noteworthy that several animal transcription factors of the homeodomain family can traffic between cells in culture using unconventional secretion and uptake mechanisms.

Meristems are the youngest tissues found in a plant, and the PD found there are thought to have the largest size exclusion limit. So far, most examples of endogenous transcription factor movement are from meristematic tissues. However, not all transcription factors with non-cell-autonomous effects move in the meristem. The latter group includes the Arabidopsis MADS domain protein APETALA3 (AP3), the ortholog of DEF, and its heterodimerisation partner PISTTLLATA (PI), both of which are similar in size to GFP.

Although it is reassuring that not all transcription factors move freely in the plant, these observations indicate that there must be active mechanisms that regulate transcription factor movement. So far, it has been largely assumed that non-movement is the default, and that movement is predominantly active. An alternative hypothesis would be that movement by diffusion is a default pathway. Movement by diffusion would be consistent with the graded distribution of LFY protein observed in plants that expressed LFY RNA only in the L1. Even if movement is at least partially by diffusion, this would not diminish the biological significance of intercellular protein trafficking. In a scenario where diffusion is an important mechanism for intercellular protein movement, retention of signalling molecules and transcription factors in their source tissue may be as important as active transport.

Retention could be achieved by targeting proteins to specific cellular compartments, or by complexing a protein with other factors. The latter scenario could apply in the case of AP3 and PI, for they are known to form stable complexes with other MADS proteins. In the case of SHR, the hypothetical interaction partner (s) might only be present in the endodermis, and effectively trap SHR (although one would have to postulate that this trapping mechanism is somehow defeated when the amount and/or timing of SHR production is changed in SCR::SHR transgenic lines.

11

Continuity in Cells

Cell Continuity is the unbroken succession of cells since life evolved 3.8 billion years ago. New cells can only be produced by cell division. New cells are needed for:

— reproduction,

— formation of multicellular organisms, and

— cell replacement.

There are three stages of the cell cycle:

— *Interphase:* The cell grows, proteins are made, the number of organelles increases, DNA replication.

Division of the Nucleus:

— *Mitosis:* two daughter nuclei, genetically identical the original nucleus, are formed.

— *Meios:* formation of four haploid genetically different daughter nuclei from the original diploid nucleus.

Cytokinesis

— The cytoplasm divides between the new daughter nuclei.

— Therefore each nucleus with its allocation of cytop becomes a new cell.

Reproduction: An emergent property associated with life. The fact that organisms reproduce their own kind is a consequence of heredity.

— Results from the transmission of heredity units (genes) from parents to offspring

— Because they share similar genes, offspring more closely resemble their parents or close relatives than unrelated individuals of the same species.

Variation: Inherited differences among individuals of the same species

— Offspring also diverge from parents as a consequence of inherited differences among them

— Genetics as a science has increased our understanding about the mechanisms of variation and heredity.

Heredity

Inheriting Chromosomes

— *DNA:* type of nucleic acid that is a polymer of four different kinds of nucleotides (ATCG)

— *Genes:* units of hereditary information that are made of DNA and are located on chromosomes.

— Have specific sequences of nucleotides, the monomers of DNA

— Most genes programme cells to synthesise specific proteins; the actions of these proteins produce an organism's inherited traits.

Inheritance is Possible because:

— DNA is precisely replicated producing copies of genes that can be passed along from parents to offspring

— Sperm and ova carrying each parent's genes are combined in the nucleus of the fertilized egg.

Transmission of genes from parents to offspring depends on the behaviour of chromosomes (organisation unit of heredity material in the nucleus of eukaryotic organisms)

— Consist of a single long DNA molecule (double helix) that is highly folded and coiled along with proteins
— Contain genetic information arranged in a linear sequence
— Contain hundreds or thousands of genes, each of which is a specific region of the DNA molecule, or locus (specific location on a chromosome that contains a gene; plural = loci)

Asexual and Sexual Reproduction

Asexual Reproduction	*Sexual Reproduction*
Single individual is the sole parent	Two parents give rise to offspring
Single parent passes on all its genes, its offspring.	Each parent passes on half its genes to its offspring.
Offspring are genetically identical to the parent,	Offspring have a unique combination of genes inherited from both parents.
	Results in a clone, or genetically
identical individual. Rarely, genetic differences occur as a result of mutation, a change in DNA.	Results in greater genetic variation; offspring vary genetically from their siblings and parents.

Understanding how DNA codes for the structural proteins that make up living things, and how it specifies the enzymes that govern biochemical processes requires a working knowledge of the structure of these information molecules and how they are replicated from one generation of cells to the next. Prokaryotic cells (bacteria) have all of their several thousand genes distributed along single long molecule of DNA.

Eukaryotic cells (protists, plants and animals) have multiple DNA molecules (multiple chromosomes). When a cell divides each daughter cell receives a complete set of chromosomes. This is accomplished through a process of chromosome sorting and distribution known as mitosis. Genetic variation is generated during sexual reproduction via meiosis.

Meiosis in Sexual Life Cycles

Fertilization: Meiosis and fertilization result in alternation between the haploid and diploid condition

Life Cycle: sequence of stages in an organism's reproductive history, from conception to production of its own offspring

Somatic Cell: any cell other than a sperm or egg cell

- — Human somatic cells contain 46 chromosomes distinguishable by differences in size, position of the centromere, and staining or banding pattern.
- — Using these criteria, chromosomes from a photomicrograph can be matched into homologous pairs and arranged in a standard sequence to produce a karyotype (a display or photomicrograph of an individual's somatic-cell metaphase chromosomes that are arranged in a standard sequence.
- — Karyotype can be used to screen for chromosomal abnormalities

Homologous Chromosomes (Homologues): a pair of chromosomes that have the same size, centromere position, and staining pattern.

- — Homologous autosomes (a chromosome that is not a sex chromosome) carry the same genetic loci; however, humans sex chromosomes carry different loci even though they pair during prophase of meiosis I.

Sex chromosomes: dissimilar chromosomes that determine an individual's sex

- — Females have a homologous pair of X chromosomes.
- — Males have one X and one Y chromosome.

— Thus, humans have 22 pairs of autosomes and one pair of sex chromosomes.

Chromosomal pairs in the human karyotype are a result of our sexual origins.

— One homologue is inherited from each parent.

— Thus, the 46 somatic cell chromosomes are actually two sets of 23 chromosomes; one a maternal set and the other a paternal set.

— Somatic cells in humans and most other animals are diploid (condition in which cells contain two sets of chromosomes (2n)

Haploid: condition in which cells contain one set of chromosomes; it is the chromosome number of gametes (a haploid reproductive cell) and is abbreviated *n*

— Sperm cells and ova are gametes, and they differ from somatic cells in their chromosome number. Gametes only have one set of chromosomes.

— Human gametes contains single set of 22 autosomes and one sex chromosome (either an X or Y)

— Thus, the haploid number of humans is 23.

The diploid number is restored when two haploid gametes unite in the process of fertilization (the union of two gametes to form a zygote).

Zygote: a diploid cell that results from the union of two haploid gametes

— Contains the maternal and paternal haploid sets of chromosomes from the gametes and is 2n

— As humans develop from a zygote to sexually mature adults, the zygote's genetic information is passed with precision to all somatic cells by mitosis.

Gametes are the only cells in the body that are not produced by, mitosis but are produced in the ovaries or testes by the process of meiosis (a special type of cell division that produces haploid cells and compensates for the doubling of chromosome number that occurs at fertilization).

— Meiosis in humans produces sperm cells and ova, which contain 23 chromosomes.

— When fertilization occurs, the diploid condition (2n=46) is restored in the zygote.

Alternation of meiosis and fertilization is common to all sexually reproducing organisms; however, the time of these two events in the life cycle varies among species. There are three basic patterns of sexual life cycles.

Animals

Gametes are the only Haploid Cells

— Meiosis occurs during gamete production. The resulting gametes undergo no further cell division during fertilization.

— Fertilization produces a diploid zygote that divides by mitosis to produce a diploid multicellular animal.

Fungi and Some Protists

In many, the only diploid stage is the zygote.

— Meiosis occurs immediately after the zygote forms.

— Resulting haploid cells divide by mitosis to produce a haploid multicellular organism.

— Gametes are produced by mitosis from the already haploid organism.

Plants and Some Algae

Alternate between multicellular haploid and diploid generations (called alternation of generations).

— The multicellular diploid stage is called a sporophyte (spore-producing plant). Meiosis in this stage produces haploid cells called spores.

— Haploid spores divide mitotically to generate a multicellular haploid stage called a gametophyte, or gamete-producing plant.
— Haploid gametophytes produce gametes by mitosis.
— Fertilization produces a diploid zygote which develops into the next sporophyte generation.

Reduces Chromosome front Diploid to Haploid

Meiosis and sexual reproduction significantly contribute to genetic variation among offspring. Meiosis includes steps that closely resemble corresponding steps in mitosis

— Like mitosis, meiosis is preceded by replication of the chromosomes.
— Meiosis differs from mitosis in that this single replication is followed by two consecutive cell divisions: meiosis I and meiosis II
— These cell division produce four daughter cells instead of two as in mitosis.
— The resulting daughter cells have half the number of chromosomes as the original cell; whereas, daughter cells of mitosis have the same number of chromosomes as the parent cell.

Meiotic Cell Division

Interphase I:

— Chromosomes replicate (same as mitosis)
— Each duplicated chromosome consists of two identical sister chromatids attached at their centromeres.
— Centriole pairs in animal cells also replicate into two pairs.

Meiosis 1.

Segregates the two chromosomes of each homologous pair and reduces the chromosome number by one-half.

It includes the following four phases:

- *Prophase 1* (longer and more complex process than mitosis).
- Chromosomes condense
- Synapsis occurs. During this process, homologous chromosomes come together as pairs.
- Chromosomes condense further until they are distinct structures that can be seen with a microscope. Since each chromosome has two chromatids, each homologous pair appears as a complex of four chromatids or a tetrad.
- In each tetrad, sister chromatids of the same chromosome are attached chiasmata, sites where homologous strand exchange or crossing over occurs.
- Chromosomes thicken further and detach from the nuclear envelope.

The cell prepares for nuclear division:

- Centriole pairs move apart and spindle microtubules form between them.
- Nuclear envelope and nucleoli disperse.
- Chromosomes begin moving to the metaphase plate
- Prophase I occupies more than 90 per cent of the time required for meiosis.
- *Metaphase I:* Tetrads are aligned on the metaphase plate.
- Each synaptic pair is aligned so that centromeres of homologues point towards opposite poles.
- Each homologue is thus attached to kinetochore microtubules emerging from the pole it faces, so that the two homologues are destined to separate in anaphase and move towards opposite poles.
- *Anaphase I:* Homologues separate and are moved towards the poles by the spindle apparatus.
- Sister chromatids remain attached at their centromeres and move as a unit towards the same pole, while the homologue moves towards the opposite pole.

- This differs from mitosis during which chromosomes line up individually on the metaphase plate (rather than in pairs) and sister chromatids are moved apart towards opposite poles of the cell.
- *Telophase I and cytokinesis:* The spindle apparatus continues to separate homologous chromosome pairs until the chromosomes reach the poles.
- Each pole has a haploid set of chromosomes that are each still composed of two sister chromatids attached at the centromere.
- Usually, cytokinesis occurs simultaneously with telophase I, forming two haploid daughter cells. Cleavage furrows form in animal cells, and cell plates form in plant cells.
- In some species, nuclear membranes and nucleoli reappear, and the cell enters a period of interkinesis before meiosis II. In other species, the daughter cells immediately prepare for meiosis II.
- Regardless of whether a cell enters interkinesis, no DNA replication occurs before meiosis II.

Meiosis II

This second meiotic division separates sister chromatids of each chromosome.

- Prophase II.
- If the cell entered interkinesis, the nuclear envelope and nucleoli disperse.
- Spindle apparatus forms and chromosomes move towards the metaphase II plate. Metaphase II Chromosomes align singly on the metaphase plate.
- Kinetochores of sister chromatids point towards opposite poles.
- Anaphase II
- Centromeres of sister chromatids separate.
- Sister chromatids of each pair (now individual chromosomes) move towards opposite poles of the cell.

— Telophase II and Cytokinesis
— Nuclei form at opposite poles of the cell.
— Cytokinesis occurs producing four haploid daughter cells.

Mitosis

In eukaryotic cells, the DNA is coiled around histone proteins forming units called nucleosomes. Nucleosomes are spaced along the DNA like beads on a string, a packaging strategy that shortens the total length of DNA to one-sixth of its uncoiled length.

At the onset of mitosis the nucleosomes become ever more tightly packed through looping and coiling to become part of a chromosome. At this stage, a chromosome is a double structure. Two identical DNA molecules produced through replication have condensed into two chromatids connected by a bridge called the "centromere." Coiling and supercoiling produces chromatids that are about a thousand times shorter than the total length of the DNA they contain.

Onion roots grow about five millimetres a day, requiring prodigious cell division. A longitudinal root section shows many cells in various stages of mitosis. In interphase the stained chromatin is loosely distributed throughout the nucleus. This is the stage where genes are being transcribed and translated into proteins. The two dark-stained bodies seen in onion root cells are nucleoli, manufacturing centres for the special RNA molecules that make up ribosomes, the machines where proteins are manufactured.

In prophase, chromosomes become clearly visible. During prophase, the nuclear envelope breaks down and disappears. The nucleoli vanish and a basket called the spindle forms around the chromosomes. In metaphase, the chromosomes line up, pulled by the spindle fibres.

During anaphase, spindle fibres pull the chromatids to opposite sides of the cell. In the final stage, telophase, a nuclear envelope forms around each set of chromosomes, and the nucleoli reappear. Mitosis has parcelled out the identical sets of genetic material, and the cell can now divide. This describes mitosis in plant and animal cells. Fungi differ in some details, as do protists.

Cytokinesis refers to the division of one cell into two daughter cells. The process varies in different evolutionary lines of eukaryotes. In animal cells, a band of actin squeezes the cell in two. Plant cells are surrounded by a rigid cell wall, requiring a different strategy. When a plant cell divides, islands of membrane form across the middle of the cell (dividing the two nuclei that resulted from mitosis) creating a partition on which a new cell wall is laid down.

Produces Genetic Variety: Meiosis is basically a process of converting the double sets of chromosomes in body cells to the single sets of chromosomes found in sex cells. At the beginning of meiosis, homologous chromosomes, one from each parent, pair up. During this process, parts of each chromatid may break off and exchange, a process called "crossing over." The next stage of meiosis reduces the chromosomes to a single set. Following reduction, the chromatids separate creating four gametes. Crossing over, and the random shuffling of parental chromosomes during reduction, assures that each gamete will contain a unique mix of genetic information.

Nucleotide Structure and Bonding: DNA is an extremely long molecule composed of two strands wound in a double helix. Each strand is made up of a chain of nucleotides. The four nucleotides found in DNA have a common structure: a phosphate group and a sugar (common to all four), and a nitrogenous base unit, of which there are four kinds: thymine, guanine, cytosine and adenine.

Two kinds of bonds make the DNA molecule possible. The backbone of each strand is formed by strong covalent bonds between the phosphate and sugar groups. These phosphate/ sugar units created the sides of the DNA ladder. The strands are held together by weak hydrogen bonds between the opposing nitrogenous bases (the rungs of the DNA ladder).

In these cross-links, thymine bonds exclusively with adenine and guanine only with cytosine,"T" with "A" and "G" with "C." The reason (as shown graphically in the programmes) is that adenine and thymine have two hydrogen bonding sites so they bond only with each other. Cytosine and guanine have three, making them exclusive bonding partners. This bonding pattern allows DNA to be replicated with a high degree of precision.

Replication of DNA: For life to have any continuity from one generation of cells to the next, the genetic blueprint must be duplicated with great precision. The beauty of the double helix is that the genetic instructions are encrypted on each strand. Therefore, each strand can serve as a template for creating its missing half so that two identical DNA molecules can be assembled from one. Essentially that is how genetic continuity is maintained through generations of cells, and through generations of individuals.

The process of DNA replication was first worked out in bacteria. Replication begins when an initiator protein chemically recognises a particular string of nucleotides (the replication origin). In the next step, enzymes known as helicases break the bonds connecting the two strands. Replication will proceed in both directions away from the replication origin.

During this process, special proteins keep the two strands apart and prevent kinking. Another enzyme, primase, lays down a primer, which serves as the starting point for a new double helix. With the DNA split, each strand can instruct the synthesis of its missing half, but to do so rapidly requires an enzyme that tracks along the strand, matching the appropriate nucleotides to create the complimentary strand. The enzyme is DNA polymerase.

Leading and Lagging Strands: During replication, double-stranded DNA is opened in a loop called a replication bubble.

At each end of the bubble is a replication fork where enzymes unravel the double helix exposing single strands of parent DNA. In the DNA double helix the backbones of the strands run in opposite directions. DNA polymerase can only build a new complementary strand by adding nucleotides in one direction. Therefore, only one strand can be replicated in a continuous sweep. This strand is called the "leading strand."

The other strand, replicated in the opposite direction is called the "lagging strand." Replicating the lagging strand requires priming and a DNA polymerase that builds lagging strand moving away from the replication fork. This produces short segments of DNA called "Okazaki fragments." Finally, the enzyme ligase welds the Okazaki fragments into a continuous strand of DNA. Replication occurs at many points simultaneously producing numerous replication bubbles that grow towards each other and merge replicating the cell's entire complement of DNA.

Twisting Problem: As the DNA is opened at a replication fork a molecular twisting problem can be imagined. As the twisted molecule unwinds from the centre, the twist tightens, preventing further opening of the helix. During DNA replication a special enzyme, topoisomerase, temporarily breaks the DNA, making it possible for the helix to rotate removing the excess twist. Releasing the twist allows the replication fork to advance.

Proofreading and Repair: In the whir of DNA replication nucleotide-based pairs must be matched precisely. But about once in every 10,000 or so pairings a mismatched nucleotide gets jammed into the new strand. When an incorrect match occurs the DNA polymerase enzyme backs up and replaces it with the correct nucleotide.

Radiation can corrupt information molecules. For example, ultraviolet radiation can rearrange the chemical bonds in adjacent thymines causing them to join together producing a "thymine dimer." A specific repair enzyme monitors DNA for thymine dimers, cutting them out, allowing DNA polymerase to correct the damaged section. In this way the genetic blueprint for each cell is passed on with very high accuracy. However, tiny errors creep in, and well they should. These mutations create the raw material for adaptation, and ultimately for the ongoing processes of evolution.

Comparison of Mitosis and Meiosis

— *Meiosis is a Reduction Division.* Cells produced by mitosis have the same number of chromosomes as the original cell, whereas cells produced by meiosis have half the number of chromosomes as the parent cell.

— *Meiosis Creates Genetic Variation.* Mitosis produces two daughter cells genetically identical to the parent cell and to each other. Meiosis produces four daughter cells genetically different from the parent cell and from each other.

— *Meiosis is Two Successive Nuclear Divisions.* Mitosis is characterised by one nuclear division.

Comparison of Meiosis I and Mitosis

	Meiosis I	*Mitosis*
Prophase	Synapsis occurs to form tetrads. Chaiasmata appear as evidence that crossing over has occurred.	Neither synapsis nor crossing over occurs.
Metaphase	Homologous pairs (tetrads) align on the metaphase plate	Individual chromosomes align on the metaphase plate.
Anaphase	Meiosis I separates pairs of chromosomes. Centromeres do not divide and sister chromatids stay together. Sister chromatids of each chromosome move to the same pole of the cell; only the homologues separate.	Mitosis separates sister chromatids of individual chromosomes. Centromeres divide and sister chromatids move to opposite poles of the cell.

Meiosis II is virtually identical in mechanism to mitosis, separating sister chromatids.

Origins of Genetic Variation

Sexual life cycles produce genetic variation among offspring

Meiosis and fertilization are the primary sources of genetic variation in sexually reproducing organisms. Sexual reproduction contributes to genetic variation by:

— Independent assortment

— Crossing over during prophase I of meiosis

— Random fusion of gametes during fertilization

Independent assortment of chromosomes: At metaphase I, each homologous pair of chromosomes aligns on the metaphase plate. Each pair consists of one maternal and one paternal chromosome.

— The orientation of the homologous pair to the poles is random, so there is a 50-50 chance that a particular daughter cell produced by meiosis I will receive the maternal chromosome of a homologous pair, and a 50-50 chance that it will receive the paternal chromosome.

— Each homologous pair of chromosomes orients independently of the other pairs at metaphase I; thus, the first meiotic division results in independent assortment of maternal and paternal chromosomes.

— A gamete produced by meiosis contains just one of all the possible combinations of maternal and paternal chromosomes.

Independent assortment: The random distribution of maternal and paternal homologues to the gametes or, more specifically, the random distribution of genes located on different chromosomes.

— Since each homologous pair assorts independently from all the others, the process produces 2n possible combinations of maternal and paternal chromosomes in gametes, where n is the haploid number.

— In humans, the possible combinations would be 223, or about eight million.

— Thus, each human gamete contains one of eight million possible assortments of chromosomes inherited from that person's mother and father.

— Genetic variation results from this reshuffling of chromosomes, because the maternal and paternal homologues will carry different genetic information at many of their corresponding loci.

Crossing over: Another mechanism that increases genetic variation is the process of crossing over, during which homologous chromosomes exchange genes.

Crossing over — the exchange of genetic material between homologues; occurs during prophase of meiosis I. This process:

— occurs when homologous portions of two non-sister chromatids trade places. During prophase I, X-shaped chiasmata become visible at places where this homologous strand exchange occurs.

— produces chromosomes that contain genes from both parents.

— in humans, there is an average of two or three crossovers per chromosome pair.

— synapsis during prophase I is precise, so that homologues align gene by gene. The exact mechanism of synapsis is still unknown, but involves a protein apparatus, the synaptonemal complex, that joins the chromosomes closely together.

Random fertilization: Random fertilization is another source of genetic variation in offspring

— In humans, when individual ovum representative of one of eight million possible chromosome combinations is fertilized by a sperm cell, which also represents one of eight million possibilities, the resulting zygote can have one of 64 trillion possible diploid combinations.

Evolutionary Adaptation Depend on a Population's Genetic Variation

Heritable variation is the basis for Charles Darwin's theory that natural selection is the mechanism for evolutionary change. Natural selection:

— Increases the frequency of heritable variations that favour the reproductive success of some individuals over others

— Results in adaptation, the accumulation of heritable variations that are favoured by the environment.

— In the face of environmental change, genetic variation increases the likelihood that some individuals in a population will have heritable variations that help them cope with the new conditions.

There are two sources of Genetic Variation:

— Sexual reproduction. Results from independent assortment in meiosis I, crossing over in prophase of meiosis I, and random fusion of gametes during fertilization.

— Mutation, which are random and relatively rare structural changes made during DNA replication in a gene could result from mistakes.

12

Cellular Descendents (Meiotic Divisions)

The sperm of angiosperms are cellular descendants of meiotic divisions occurring in the anther of the flower. Meiosis in each meiocyte forms four genetically dissimilar microspores. Each microspore undergoes a mitotic division to form a large vegetative cell and a smaller, reproductive generative cell. Together these cells form the pollen grain or, strictly speaking, the microgametophyte. Subsequently, the generative cell migrates into the vegetative cell and becomes an elongated, spindle-shaped, or even filiform cell, depending on the species; each generative cell retains its own plasma membrane and is, in turn, enveloped by the vegetative cell membrane.

The generative cell undergoes a second mitotic division to form the two sperm. Depending on the timing of generative cell mitosis the two sperm cells may be formed before germination of the pollen (tricellular) or after pollen germination (bicellular). Of 243 families surveyed, bicellular pollen was present in 137 families (56 per cent), tricellular pollen in

55 families (23 per cent), and both types of pollen in 51 families (21 per cent). Bicellular pollen is typically regarded as the ancestral condition in most families, tricellular pollen as the derived condition. Sperm cells isolated from pollen grains in tricellular species. Bicellular species must be cultured in order to grow pollen tubes, trigger mitosis, and obtain sperm cells.

Development of Sperm Cells

Characterisation of Plant Sperm Cells: Flowering plant sperm are structurally simple, apparently non-motile cells that contain a normal complement of cellular organelles, including heritable cytoplasmic organelles such as mitochondria and, in some plants, plastids. In contrast to animal species, competition in angiosperms is a property of the gametophyte (pollen grain and tube) and not the gamete; once this competition is completed, both sperm cells contribute to the formation of the embryo during double fertilization: One sperm cell fuses with the egg to produce the embryo, the other fuses with the central cell to produce the nutritive endosperm.

The sperm cells are typically connected to one another and physically associated with the vegetative nucleus, transported within the pollen tube as a "linked unit" termed the "male germ unit". The linkage of the sperm cells increases the effectiveness of gamete delivery, reduces hetero-fertilization, and ensures nearly simultaneous transmission of the two sperm cells into the receptive embryo sac. This association is present in most species flowering plants studied to date or is formed soon after pollen germination.

Preferential Fertilization and Sperm Dimorphism: Although the association between one sperm cell and the vegetative nucleus imposes a degree of polarity, strong cellular differences may also occur between the two sperm, including differences in the content of heritable organelles termed cytoplasmic heterospermy, and differences in the nucleus termed nuclear heterospermy. Although cytoplasmic heterospermy may be relatively common in flowering plants nuclear differences have been reported only in Zea mays, in which B-chromosomes frequently undergo non-disjunction during generative cell mitosis.

The most common form cytoplasmic dimorphism is mitochondrial inequality: Sperm cells associated with the vegetative nucleus usually contain more mitochondria than the other cell. The most extreme form examined to date is found in Plumbago zeylanica, where both plastid and mitochondrial content differ. Preferential fertilization, in which one sperm cell has a greater likelihood of fusing with the egg, may also occur in species where the sperm cells differ.

In Plumbago, fusion between the egg and the plastid-rich sperm cell occurs in 94 per cent of the cases examined, selectively transmitting male plastids into the zygote. Transmission of male mitochondria into both the egg and central cell, however, occurs at a nearly constant 1:1000 ratio of male:female mitochondria, because the differences in mitochondrial content in the egg and central cell match those of the dimorphic sperm cells. Whether this ratio represents the maximum permissible dose of sperm mitochondria to prevent sperm transmission of mitochondrial DNA or the minimum needed to make recombination of mitochondrial DNA possible is unclear.

Preferential fertilization (termed meiotic drive by geneticists) also occurs Zea, in which the sperm cell containing one set of extra B-chromosomes fuses with the egg about 65 per cent of the time. Rather than simply providing marker for identifying the sperm cell destined to fuse with the egg, however, B-chromosomes appear to confer a selective advantage to the sperm cell that contains them. It is interesting that preferentiality may be eliminated introducing excessive B-chromosomes into the sperm or by introducing a specific B-chromosome (TB-9b) into the egg.

This observation suggests that maternal discrimination is active only under relatively strict conditions. Maternal control of fertilization is also suggested in a mutant line of barley in which seeds are produced containing normal endosperm but haploid embryos. One sperm cell apparently fuses with the central cell regardless of pollen source whereas the other remains unfused. Research on the mechanism of fertilization in normal lines of barley indicates that the zygote of plants with uniparental; maternal cytoplasmic inheritance are cytoplasmically restrictive: Male cytoplasmic organelles may be shed by the separation or pinching off of cellular folds during maturation of the sperm and the sperm cytoplasm excluded from the egg, remaining outside of the cell.

However, the central cell in the same plant may be cytoplasmically permissive, allowing sperm organelles to be transmitted into the central cell. Plants with biparental inheritance appear to be cytoplasmically permissive during the entire double fertilization process.

Sperm Cell Isolation

Sperm cell isolation was first reported in detail by Cass, who placed pollen grains of barley in a Brewbaker-Kwack (BK) pollen germination medium with 20 per cent sucrose, obtaining release of the sperm cells by osmotic shock and observing the behaviour of the living cells using Nomarski differential interference contrast microscopy. Over 30 min, the originally paired sperm cells separated, lost their spindle shape, and became ellipsoidal to spheroidal. Alternative media, such as filtrates of ovular material, were also examined but there were no specific benefits or effects on sperm cell appearance or behaviour.

Numerous isolation protocols have emerged including some modified for collection of sperm cells en masse. Sperm cell release has been conducted using two general techniques: osmotic shock, or direct physical separation by grinding or applied pressure. The pollen cytoplasm and gametes are usually incubated for an additional 10-20 min to allow the gametes to separate from vegetative cell cytoplasm, and then pollen cell walls are removed by filtration.

A nylon mesh two thirds to one half the average diameter of the pollen efficiently captures the pollen walls. En masse isolation of sperm cells involves either centrifugation on a discontinuous gradient or filtration using a polycarbonate (e.g. Nuclepore) filter. Isolation of sperm cells using osmotic shock relies on the greater sensitivity of the vegetative cell membrane to changes in the milieu. In some cases, the shock of entering an aqueous medium from a nearly desiccated state is sufficient to rupture pollen grains regardless of the osmotic concentration used, but sperm cells may be surprisingly intact. After pre-hydration (example, exposure in a humid chamber), pollen is more gradually introduced to a hydrated state, possibly enhancing viability of the isolated sperm cells. Pollen grains may then be subjected to an abrupt change in osmotic concentration or pH or lysed in other ways.

The isolation of sperm cells can also be accomplished by using physical stress to break the pollen plasma membrane. Methods used include glass tissue homogenisers or the pressure of a glass roller applied to smooth glass surface. Clearance height in the homogeniser is not critical, because the resistant exines will set the minimum. Isotonic or slightly hypertonic media are best for the isolation of the cells.

The most frequently used media for sperm isolation are modifications of the BK medium (1)—e.g. 100 ppm H_3BO_3, 300 ppm $Ca(NO_3)_2 . 4H_20$, 200 ppm $MgSO_4 . 7H_2O$, 100 ppm KNO_3, and 10 per cent sucrose. Other successfully applied media include the Roberts medium, which is a Tris-buffered modification of BK medium; a modified K3 Kao protoplast culture medium containing BK macronutrients, plus micronutrients, co-enzymes, vitamins, and hormones; and the RY-2 protoplast medium used for plantlet regeneration. Because few of the cited reports use parallel methodology, it is difficult to conclude which, if any, of the media is ideal for most angiosperms; but the life expectancy of isolated sperm cells seems strongly improved by complex media and particularly by the addition of vitamins protocols are refined, a distinction will likely be made between isolation media (designed to optimise recovery) and storage or culture media (designed to extend sperm cell viability).

Assessment of Sperm Cell Quality

Sperm cells have been assessed for:

(a) structural intactness,
(b) membrane integrity,
(c) enzyme activity, and
(d) respiratory activity.

All of these aspects are appropriate; perhaps most crucial would be an assay of the cell's ability to effect fertilization, which is not currently feasible. Techniques to assay structural intactness include transmission and scanning electron microscopy, which can be used assess membrane structure and cell shape; intramembrane particle distribution can be inferred from freeze-fracture replicas. Integrity of membranes can be assessed using the exclusion of polar dyes such as Evans blue propidium iodide, both of which are rapidly absorbed into dead cells.

Enzyme activity is readily demonstrated using the so-called fluorochromatic reaction in which fluorescein diactate is cleaved cellular esterase. Fluorescein diacetate is easily absorbed by living cells but becomes fluorescent only if esters are cleaved from the fluorescein molecule and the cell is intact enough to retain the polar dye. Respiratory activity has been demonstrated using an assay for ATP involving cleavage by luciferin yielding measurable luminescence. The results of this method roughly correspond to those using fluorescein diacetate but ATP is still evident even when the cells are no longer viable according to the previous method.

During isolation, the sperm cells commonly assume an ellipsoidal to spheroidal shape and lose their microtubules.

A comparable loss microtubules has been described in detail using anti-tubulin immunofluorescence of freshly isolated generative cells in bicellular plants. These changes appear comparable to those observed in isolated generative cells. An exception to this is the isolation of sperm cells in Gerbera, in which the native shape is retained. Spheroidal sperm cells are not typically seen in young growing pollen tubes in vivo, but a significant rounding of these cells seems characteristic as pollen tubes near the ovule. Although spheroidal sperm cells obtained through isolation are artifactual, this feature is common in somatic plant protoplasts, typically without impairing cell quality.

Isolated sperm cells lose the plasma membrane of the vegetative cell that usually surrounds them in vivo. This exposes the true surface of the sperm cells and makes it possible to characterise the plasma membrane. In vivo, the same modification of the surface of the sperm cell occurs once sperm cells are discharged into the embryo sac and postulated to be a prerequisite for sperm fusion in angiosperms. A significant remaining question about the condition of isolated sperm cells is whether they are completely mature in the pollen grain or whether essential gene products remain to be transcribed or translated during pollen tube growth prior to fertilization. Given the vast range of times between pollination and gametic fusion in different angiosperms, these plants likely vary in this respect.

Psychological Characterisation of Sperm Cells

Immunological Characterisation: Hybridomantibodies have been elicited to an inoculum of intact sperm cells of Brassica and Plumbago to generate antibody libraries to surface epitopes. Intact cells were used because of the small number of sperm cells available and the technical problems of preparing isolated plasma membranes. Depending on how long sperm cells remain intact in the inoculum, whole cells would be likely to elicit some antibodies to surface compounds.

Initial attempts were partially successful, but techniques for retaining integrity of the sperm cells during inoculation and for increasing the sensitivity of screening for surface-specific epitopes will be needed before it is possible to produce a useful monoclonal antibody library for recognition studies. The only detailed report on antibodies elicited to isolated sperm, in Plumbago, indicates the feasibility of this approach.

Sperm antigens are effective in eliciting antibody-producing hybridoma lines to a wide spectrum of different epitopes in which pollen wall compounds, including allergens, are not immunodominant. The secreting lines were specific for epitopes in the sperm-enriched (23 per cent), cytoplasmic-particulate (4 per cent), and water-soluble fractions (8 per cent). Antibody-producing lines were mainly IgM (61 per cent), reflecting high glycoprotein content of the cells. Identification of a recognition factor from this approach is complicated by the need to develop an appropriate functional assay. Further research in this area is warranted if specific recognition compounds occur as in pollen tube-pistil interactions.

Polypeptide/Protein Characterisation: Biochemical characterisation of the proteins of male gametes using SDSPAGE has been conducted on a small number of species: Brassica, Gerbera, Plumbago, and Zea. These results indicate that unique polypeptide bands are readily identified in sperm fractions of higher plants.

In order to reduce this effect, presence/absence comparisons with possible contaminants were conducted in Plumbago. Two-dimensional gel electrophoresis (IEF/SDS-PAGE) was conducted compare polypeptide heterogeneity in P. zeylanica. The sperm-enriched fraction and two selected contaminant fractions were examined. Cytoplasmic particulates pelleted at 100,000 x *g* constituted the first fraction; water soluble molecules collected from the supernatant formed the second fraction. The sperm-enriched fraction contained the most polypeptide spots. The cytoplasmic-particulate fraction yielded 427 spots, and the water-soluble fraction had 285. One quarter of the spots were found in all fractions, about half were found in two fractions, and one quarter were found in a single fraction.

Of the latter group, 51.9 per cent of the polypeptides were unique to the sperm-enriched fraction, 40.4 per cent to the water-soluble fraction, and 2 per cent to the cytoplasmic-particulate fraction. Polypeptides of sufficient molecular weight to be involved in recognition (Mr 70,000 to >200,000) were present in the sperm-enriched fraction. Some of differences between polypeptides undoubtedly result from generational differences of gene expression between the sporophyte and gametophyte.

Although pollen wall proteins are synthesised by the sporophyte, gametophytic gene products may control gametic recognition, fusion, and fertilization. No studies of developmental polypeptide changes in the male gametic line are currently available. Polypeptide changes in developing pollen and microspores of Zea, Triticum, and Brassica indicate that developmental changes do occur. In Triticum, 11 new bands were detected in SDS-PAGE and IEF gels and 4 bands were lost during pollen maturation, indicating a small number of stage-specific changes. Changes during anther development in Zea reflect similar stage-specific changes in the proteins and activities of specific enzymes. With further research, considerable progress in identifying sperm-specific polypeptides may be expected.

Elemental Characterisation: Energy-dispersive X-ray analysis of pollen components indicates that sperm cells contain significant quantities of carbon, oxygen, phosphorus, calcium, and potassium, with smaller amounts of magnesium, silicon, manganese, and detectable levels of chromium. Although mostly similar, the dimorphic sperm of Plumbago exhibit small differences in Kα lines of calcium, potassium, and phosphorus. Other studies of pollen have indicated similar elemental composition in the pollen tube tip. The surfaces of pollen grains different species, however, display significant elemental differences.

Molecular Characterisation: Conflicting information is available regarding the synthesis of RNA in sperm cells. In the tricellular pollen of Secale cereale, [5-^{3}H] uridine was incorporated into both the sperm and vegetative nuclei during in vitro pollen tube growth. The types of RNA synthesised by the sperm nuclei were not determined. However, in the bicellular pollen of Hyoscyamus niger, exposure to short pulses of [5-^{3}H] uridine did not detect incorporation. Determination of whether RNA synthesis is related to the timing of sperm cell formation will require additional research.

Translational products within sperm cells have been obtained by means of in vitro translation of isolated RNA species using [^{35}S] methionine, combined with electrophoresis and radioautography (P. Roeckel and C. Dumas, personal communication). These studies demonstrate that translatable mRNA pools exist for a number of polypeptides in the sperm cells of corn; their identity has not yet been fully determined. The use of stage-specific promoters and reporter genes in combination with particle-gun transformation to elicit transgenic plants seems to be a promising approach to understanding the developmental processes underlying sperm differentiation.

Adaptation and application of existing molecular biological methods to sperm cell biology are critically needed.

Male-germ-cell Transcripts in Nicotiana Tabacum

Development and differentiation of plant male gametes starts with an asymmetric division of the microspore, resulting in a relatively large vegetative cell and a much smaller generative cell that after one further division will give rise to two sperm cells. In tricellular pollen like that of rape (Brassica) seed and grasses such as maize (Zea) and wheat (Triticum), the division of the generative cell occurs in the pollen grain before maturation.

In bicellular pollen like that of lily (Lilium) and tobacco (Nicotiana), the division of the generative cell occurs after dispersal of the pollen and its germination on the surface of the stigma. The switch from vegetative to generative development relates to the asymmetry of the microspore division, but the actual triggers and regulation are still unknown. The development of the large vegetative cell is accompanied by the expression of a number of specific genes such as LAT52 and LAT59, Zml3, and NTP303 that may relate to pollen maturation, germination and pollen tube growth.

In contrast, the cytoplasmic volume of the generative cell is extremely small and its nucleus becomes heterochromatic, indicating little or no gene activity. With the exception of LGC1 from lily, no specifically expressed genes have been isolated. Although the symmetric division of the generative cell produces two sperm cells that may differ in size and morphology, the sperm cells remain connected as a single male germ unit. The cytoplasmic volume of the sperm cells always stays very small and the nuclei remain heterochromatic. At fertilization, one of the two sperm cells fuses with the egg cell that then will develop into the embryo, while the other sperm cell fuses with the central cell that will form the endosperm.

Until now, only the morphologically different sperm cells of Plumbago zeylanica have been shown to fuse preferentially with either the egg or the central cell; sperm cells from all other species examined to date have not shown this preferential fusion. Therefore, it is still unknown whether a general mechanism that targets sperm cells specifically to the egg and central cell exists, and how this mechanism might work. Due to their inaccessibility and low transcriptional activity, little is known about the molecular biology of the male gamete.

For example, it is not known whether the gene expression programme is different between sperm cells, and between the sperm cells and the generative cell. So far, screening of cDNA libraries from pollen and pollen tubes has not provided any markers for male gamete development. However, recently developed procedures for mass isolation of generative cells and sperm cells enabled the isolation and characterisation of sperm-cell-expressed genes, such as the male-gametespecific gene LGC1 of lily, as well as the identification of sperm plasma membrane proteins of Brassica campestris, Lilium longiflorum, and Zea mays.

Plants of N. tabacum L. cv. 'Petit Havana' were grown in soil under standard greenhouse conditions. Flower tissues were collected at different stages according to the morphological markers. Mature pollen was collected from flowers at stage 12 and petals, ovaries, anthers, and pistils without ovaries were collected at stage 10 unless otherwise indicated. For pollination experiments, flowers were emasculated at stage 11 and pollinated with fresh pollen 1 day later. Pollen tubes were cultured in vitro according to Read et al., washed in 15 per cent sucrose prior to collection, frozen in liquid nitrogen, and stored at -80°C.

Isolation and Purification of Sperm Cells: Pollen tubes were cultured in vitro for 22 h, collected on 30 urn nylon filters and washed in 15 per cent sucrose to remove the growth medium. To release the sperm cells, the pollen tubes were incubated in a cell-wall-degrading enzyme mixture [0.6 per cent Cellulase and 0.3 per cent Macerozyme (Yakult/Honsha) in 2 mM MES buffer (pH 6.7), 8 per cent sucrose, and 0.1 per cent BSA] for 20 min at room temperature. Burst pollen tubes were removed by filtration through four layers of 11 um nylon mesh and the final sucrose concentration of the filtrate was adjusted to 17 per cent by addition of 30 per cent sucrose in 2 mM MES buffer (pH 6.7). This suspension was layered onto a discontinuous Percoll gradient consisting of two layers of 7.5 ml of 15 per cent and 22 per cent Percoll in 15 per cent sucrose, 2 mM MES (pH 6.7), and centrifuged for 30 min at 2,700 g at 4°C.

Sperm cells were collected from the sperm-cell-rich layer at the bottom of the 15 per cent Percoll zone and washed twice in 15 per cent sucrose, 2 mM MES (pH 6.7). To check yield and purity, samples were analysed by microscopy under Nomarski differential interference contrast with a Zeiss Axiomat III microscope and by fluorescence microscopy after DNA staining with DAPI (AppliChem) in a Leitz Ortholux-orthoplan with appropriate filters and illumination. Samples were frozen in liquid nitrogen and stored at -80°C for RNA isolation. Sperm cells that were isolated for RNA to be used in RT-PCR were treated with 50 ug/ml RNase (Promega) at 37°C for 30 min prior to freezing.

Nucleic Acid Isolation: Quantities of 0.1-1 g of frozen tissue were homogenised in 2 ml RNA extraction buffer [0.1 M Tris (pH 8.0), 0.1 M NaCl, 1 per cent SDS, 50 mM EDTA, 50 mM *B*-mercaptoethanol, and 1 per cent triiso-propyl-naphtalene-sulphonic acid sodium salt] and protein was removed by phenol extraction. Total nucleic acids were precipitated with 0.1 volume of 3 M sodium acetate (pH 5.2) and 2.5 volumes of 100 per cent ethanol. The resulting pellet was dissolved and RNA was precipitated with one-third volume of 8 M LiCl. RNA for RT-PCR was treated with 1 U DNase for 30 min at 37°C, extracted with phenol and precipitated with ethanol. Sperm-cell poly (A^+) mRNA was purified using the Qiagen Direct mRNA Isolation Kit according to the manufacturer's protocol.

The tobacco sperm cell cDNA library was constructed following the procedure provided with the ZAP-cDNA Gigapack III Gold cloning kit (Stratagene). Sperm-cell poly(A^+) mRNA (5 ug) was used for synthesis of the first strand cDNA. Size selected cDNA molecules (1-3 kb) were ligated into Uni-ZAP XR vectors. The primary library was amplified once prior to screening. Screening for sperm-cell specific cDNAs occurred in two rounds. In the first round the cold plaque approach was adopted to identify low abundance genes.

Two duplicate plaque filters were hybridised to two different ^{32}P-labelled cDNA probes of mRNA from mature pollen and of pooled mRNA from young flowers (stage 6) and seedlings. The cDNA probes were prepared with the reverse transcriptase Superscript II kit (Gibco/BRL) using 1 ug of poly(A^+) mRNA. The hybridisation temperature was 65°C and the filters were washed with 0.1 x SSC, 1 per cent SDS at 65°C. "Cold" plaques that did not hybridise to either of the two probes were selected as possible sperm cell-specific transcripts.

In the second round, candidates were screened by reverse northern hybridisation. pBluescript phagemids were prepared from the selected cold plaques following the manufacturer's in vivo excision protocol. The plasmids were digested with EcoRI and XhoI, run on an agarose gel and analysed by Southern hybridisation to three different ^{32}P-labelled cDNA probes: (1) sperm-cell poly (A^+) mRNA, (2) pollen poly (A^+) mRNA, and (3) pooled poly (A^+) mRNA from young flowers and seedlings.

RNA Gel Blot Hybridisation: Aliquots of 0.5 μg poly (A^+) mRNA or 20 ug total RNA were denatured, fractionated on denaturing 1.2 per cent agarose gels containing formaldehyde, transferred to Hybond-N+ nylon membranes (Amersham, UK), and hybridised to 32P-labelled cDNA inserts. Hybridisation was carried out overnight at 65°C in 5x SSC, 5x Denhardt's solution, 0.5 per cent SDS and 100 ug/ml denatured sonicated herring sperm DNA. After hybridisation, the filters were washed with 0.2x SSC, 0.1 per cent SDS at 65°C and exposed- with intensifying screens - to Kodak X-OMAT film at -80°C. To verify equal loading of RNA, the filters were stripped and probed with a tobacco 25S ribosomal probe.

RT-PCR: For reverse transcription (RT), 5 μg of DNase-treated RNA was used in a 20 μl reaction according to the manufacturer's protocol (MBI Fermentas) and for each PCR amplification, 2 µl of this RT reaction was used. The PCR reaction was carried out in 50 µl with 0.5 U Taq DNA polymerase (Eurogentec), 1.2 mM $MgCl_2$ and 50 pmol of each of the two sequence specific primers with the following cycle: 3 min at 95°C, 30 cycles of 30 s at 95°C, 30 s at the optimal annealing temperatures (54.5°C for the NtS1 and 50°C for NtS2 primers), and 2 min at 72°C, followed by 7 min at 72°C.

The NtSl-specific primer set was (XH3) 5'-ACCGGCCAAAGTTGTTATTCCTC-3' and (XH4) 5'-GCATGCTGGCGGGTTATTTT-3', the NtS2-specific primer set was (HP7) 5'-AATCTGTTCTCTAGCGGGTTCT-3' and (HP8) 5'-TTTTGGGAGGAGGACAGGTAT-3', and the NTP303 specific primer set was (F1) 5'-TTGCTACTTTGCCTCTCCGT-3" and (R1) 5'-CAGGCCCTTGACGATACCG-3'.

Sequence Analysis: Sequencing was performed by the di-deoxy chain termination method using the ABI-PRISM Big Dye terminator Cycle Sequencing Ready Reaction Kit (Applied Biosystems). The sequence data were analysed with NCBI GenBank BLAST programmes and the Swiss Institute of Bioinformatics Prosite database. The GenBank accession numbers for NtS1 and NtS2 are AF248538 and AF248539, respectively.

13

Circulation of Blood

The blood vascular system includes the *heart* and *blood vessels.* In addition, there is a lymphatic system which returns lymph to the blood. Lymph is formed from blood plasma that has filtered from the blood vessels but is unable to go back directly into blood capillaries because of blood pressure. The lymph bathes the cells of the tissues directly and carries the digested foods and oxygen which the cells need. In return, the cells give to lymph their waste substances like urea and carbon dioxide.

In higher forms the circulatory system is of *closed type,* in which the blood is confined to the vessels throughout its course from the heart and back to the heart. In many invertebrates, with an *open type* of circulatory system, the blood vessels open into numerous irregular spaces or *lacunae* and blood comes directly in contact with the cells. From these intercellular spaces blood returns to the heart.

Significance of Circulation in Invertebrates

Among many lower invertebrates the role of a circulatory system is performed by other structures. The development of a separate circulatory system is first seen in Nemerteans which have a few blood vessels in which the blood is propelled by the 'milking', action of the muscular walls of the vessels. The heart as a definite propulsive organ appears in the annelids, arthropods and molluscs. These are provided with well developed vascular systems, the arthropods and molluscs (except cephalopods) having the open type. In the earthworm in which a closed system is approached, the true heart is the dorsal blood vessel. The invertebrate heart is always located dorsal to the alimentary canal.

Significance of Circulation in Vertebrates

In vertebrates the circulatory system is of closed type and consists of a heart and blood vessels (arteries, capillaries and veins). The vertebrates have chambered hearts, having two, three or four chambers. The fish heart consists of two main chambers, auricle and ventricle. It is a *single-circuit heart,* the blood passes through it only once in each complete circuit. Amphibians and most reptiles have a 3-chambered heart. Birds and mammals have the 4-chambered heart and there are two circuits, one through the lungs *(pulmonary)* and the other through the body *(systemic).* Starting with left ventricle, the blood passes through the aorta and its many divisions to all parts of the body and then into the capillaries within the tissues. Veins collect the blood from the capillaries and carry it back to the right auricle of the heart. This circuit from the left ventricle to the right auricle is the *systemic circulation.* From the right auricle the blood passes into the right ventricle, thence through the pulmonary artery to the lungs where the gaseous exchange is made. Then by a vein or veins the blood is carried to the left auricle and thence to the left ventricle. This short circuit from the right ventricle to the left auricle is the *plumonary circulation.* In this complete circulation the blood passes through the heart twice. The advantages of the 2-circuit over the 1-circuit plan are many because there is no mixing of the oxygen-rich and oxygen-poor bloods. The blood in the aorta contains relatively more oxygen, a higher metabolic rate is thus possible and a resulting higher temperature can be maintained by the animal.

Physiological Anatomy of the Blood Vessels

(a) *Arteries:* All the blood vessels leading from the heart are called the arteries, whether they contain oxygenated blood as in the aorta or deoxygenated blood as in the pulmonary artery. The thick walls of the arteries consist of a heavy layer of elastic tissue. More small branches of an artery are called *arterioles,* having a diameter of about 0.1 mm or less.

(b) *Capillaries:* The arterioles lead into a branching series of microscopic vessels known as the capillaries that have a diameter of about 7-10 *microns.* The capillary walls are thin and allow an exchange of nutrients, gases and other substances between their blood and the cells through a process of diffusion or active transport.

(c) *Veins:* The capillaries in turn unite to form small veins that join to form larger and larger veins, which ultimately carry blood back to the right side of the heart. The veins do not have an inner layer of elastic membrane found in the arteries. Some of them possess internal valves along their length to prevent the backflow of blood.

Role of Atherosclerosis and Arteriosclerosis

These are changes in the structure of the walls of the arteries, changes that ultimately impair their function. *Atherosclerosis* is considered to be the main cause of heart attack and stroke. It is caused by the deposition of lipid, usually cholesterol, in the smooth muscle cells of the artery wall. Cholesterol deposition is followed by some calcification and is topped by a fibrous cap, forming a plaque that protrudes into the lumen of the artery and cuts off blood to those parts of the body served by the diseased vessels. Although the precise role of cholesterol is uncertain, the formation of atherosclerotic plaques is accelerated by abnormally high blood cholesterol (130) and by other factors, such as cigarette smoking and high blood pressure, 180 systolic. The cumulative effects of these three major risk factors increase the likelihood of heart attack and stroke.

Arteriosclerosis is hardening of the arteries due to the deposition of calcium and fibrous tissue within their walls. This reduces the elasticity of the arteries and consequently causes increase in pressure within them. When blood reaches the brain under such high pressure it may rupture the small vessels, depriving the tissues of oxygen and resulting in a "stroke".

Physiological Anatomy and Working of the Heart: The mammalian heart consists of four chambers, two auricles and two ventricles. Typically, one large blood vessel is connected to each of these chambers. The *aorta* leaves the *left ventricle* and the branches of this vessel supply all parts of the body with arterial or oxygen-rich blood. Venous blood collected from all body regions returns through the *vena cava* into the *right auricle.* The right auricle is connected with right ventricle through the *tricupsid valve,* an opening with three flaps. This valve allows blood to flow from the auricle to ventricle but not in the reverse direction. Venous blood collected in the right ventricle leaves this chamber via the *pulmonary artery* leading to the lungs where the blood is oxygenated and arterial blood returns through the *pulmonary vein* into the left auricle. A *bicupsid or mitral valve,* equipped with two flaps, separates the left auricle from the left ventricle. The bicupsid and tricupsid valves together may be referred to as the auriculo-ventricular valves or *A-V* valves. Smaller semi-lunar valves are situated where the aorta and the pulmonary artery leave the left and right ventricle respectively.

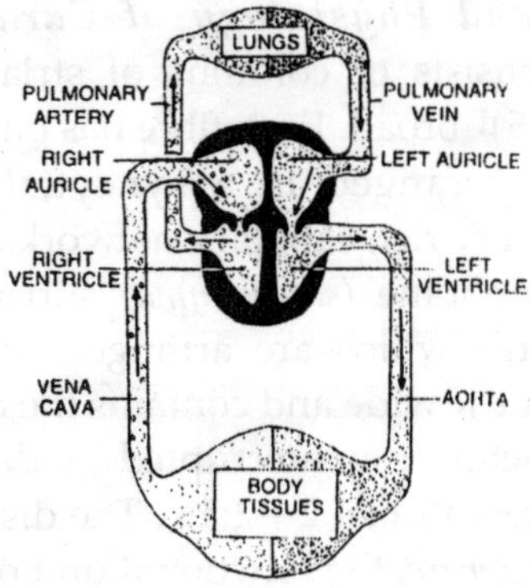

Course of blood circulation through the body.

The left chambers of the heart carrying the arterial blood are not connected directly with the right chambers, carrying the venous blood. In the mammalian heart the sinus has been incorporated into the right atrium. The wall of the heart in the adult is divided into three layers known as the endocardium, myocardium and epicardium. The *endocardium* consists of the endothelium which is continuous with that of the blood vessels and of subendothelial connective tissue. The *myocardium* is the muscle layer upon which depends the pumping action of the heart. It is thin in the auricles but very thick in the ventricles. In the left ventricle, which pumps blood into the farthest parts of the body, it is much thicker than in the right. The *epicardium* consists of the pericardial mesothelium together with underlying connective tissue. This layer is also called the *visceral pericardium* and with the *parietal pericardium* it bounds the pericardial cavity, forming a closed sac containing the pericardial fluid.

In the heart of humans and some other mammals there is a special conducting tissue concerned with the instigation and regulation of the heart beat. This specialised tissue is called *nodal tissue* and is necessary for the functional separation of the auricles from the ventricles. To produce a functional stimulation of the atrial and ventricular beats, this nodal tissue conducts impulses so rapidly to the ventricles that all their parts contract almost simultaneously.

Structure and Physiology of Cardiac Muscle: The myocardium consists of columns of striated muscle fibres, 100 μ long and 15 μ broad. Each fibre has only a single nucleus. The fibres are arranged in a syncytial fashion and are surrounded by very rich capillary network. Each fibre consists of an outer membrane *(sarcolemma)* surrounding numerous striated myofibrils which are arranged longitudinally. Each myofibril is about μ wide and contains the contractile proteins Stenosing and actin. It is interrupted by dark lines known as the Z lines at intervals of 1.2—2.5 μ. The distance between two Z lines is the *sarcomere,* the functional unit of the myocardium.

The ultrastructure of cardiac muscle is quite similar to that of skeletal muscle. The *sacromere* consists of alternate light and dark bands. The central dark band is 1.5 μ long and contains *myosin* filaments (10m μ broad). The central band, known as the A band, is flanked by an I band on each side which appears light and consists of actin filaments (5m μ broad). The sacrolemma is invaginated by a system of transverse tubules which penetrate the sacromere at the Z line. The lumina of these tubules is continuous with the extracellular fluid. In addition, there are longitudinal tubules. At each Z line these longitudinal tubules approximate to but do not anastomose with the transversely disposed invaginations of the sarcolemma. The term 'triad' is used to describe the arrangement of the terminal dilations of two longitudinal tubules (known as cisternae) one oil each side of a transverse tubule.

Depolarisation of the sacrolemma causes the development of an electric potential which is conducted through the transverse tubule and then causes depolarisation of the adjacent cisternae. This depolarisation causes a release of Ca^{2+}, which diffuses into the myofibrils and activates myosin, which then splits ATP, yielding the energy necessary for the formation of actomyosin complex and for the contraction of the muscle. The Ca^{2+} ions so released are then actively transported back into the longitudinal tubules of the sarcoplasmic reticulum and relaxation occurs.

At the ends of each fibre, or cell, is a structure known as the *intercalated disc.* Electron microscopy indicates that these discs are actually cell junctions. Thus, they are cell membranes, often arranged in a zig-zag manner, with considerable amounts of amorphous dense material associated with them. The number of *intercalated discs* seems to increase with age. Human cardiac muscle shows a clear distinction between perinuclear and peripheral cytoplasm. Between the muscle fibres is connective tissue of reticular, collagenous, and elastic fibres.

The cardiac muscle differs quantitatively from skeletal muscle in having more mitochondria crowded between the myofibrils, and in being richer in the red-pigmented myoglobin than most skeletal muscle. Myoglobin is found in the sarcoplasm of cardiac muscle fibres. It is a conjugated protein containing an iron porphyrin prosthetic group closely related to haemoglobin. Myoglobin faciliates the transport of oxygen from cell membrane to the mitochondria where tissue respiration occurs. It should be noted that the heart muscle is not a structural syncytium but it behaves as if it were, the reason being that the electrical resistance offered by the intercalated disc is very low. Thus, when one cell is excited, the excitation process spreads easily across the intercalated disc to its neighbour. This allows the impulse to spread from one fibre to another throughout the whole atrium or ventricle causing them to respond as a single unit (all-or-none principle of heart beat).

Physiologically the hearts are of two types:

1. *Myogenic Heart:* It is a heart in which the *beat* originates *myogenically, i.e.,* the heart beat is initiated and controlled by a special node of modified or unmodified heart muscles, the so-called sows *node* or *pacemaker.* Located in the wall of the right auricle, the pacemaker produces rhythmic signals leading to contraction of the heart. Because of this built-in mechanism, a myogenic heart exhibits an *intrinsic* beat. When isolated from the body and placed into an artificial nutrient medium, such a

heart may for a time contract and relax in a slow rhythm. In the body, this inherent rhythm of the heart is regulated in both, amplitude (strength) and rate by control systems, which are nervous, physical, and chemical. Acetylcholine inhibits the heart beat of a myogenic heart. The hearts of *vertebrates* and *molluscs* are myogenic.

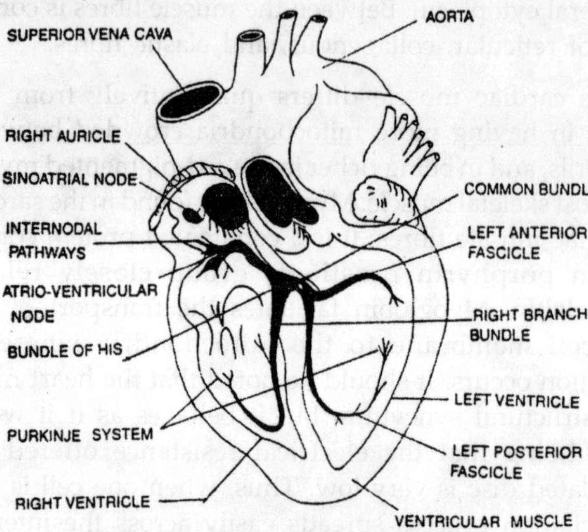

Mechanism of heart beat. Arrows indicate spread of the waves of excitation.

2. *Neurogenic Heart:* It is a heart in which the beat originates by nervous stimulation or *neurogenically, i.e.,* the contraction is initiated by the nerve ganglion situated in or close to the heart, the nerve cells acting as the pacemaker. Such a ganglion delivers rhythmic impulses to the heart muscle, and the removal of the ganglion stops the heart. Acetylcholine accelerates the heart beat rate of a neurogenic heart. The hearts of the most *arthropods* and some *annelids* are neurogenic.

Role of Heart

The cardiac conduction system consists of the following 3 elements:

1. *Sinoatrial Node:* The sinoaterial (SA) node is a specialised muscle fibre area in the posterior wall of the right atrium, below the opening of the superior vena cava. The nodal cells are smaller than the surrounding atrial cells (3 μm in diameter as compared to 5 to 11 μm for the a trial cells). The bundles of fibres from the SA node activate first the right and then the left atrium. The SA node is the normal cardiac *pacemaker* as its rate of discharge determines the rate at which the heart beats. Internodal atrial pathways exist between the SA node and the AV node.
2. *The Atrioventricular* (AV) *Node:* It is in the atrial septum, just above the tricupsid valve. The AV node is normally the only conducting pathway between the atria and ventricles.
3. *Bundle of His:* The AV node is continuous with the bundle of His, which gives off a left bundle branch at the top of the interventricular septum and continues as the right bundle branch. The left bundle branch divides into an anterior fascicle and posterior fascicle.

   The branches and fascicles pass down either side of the septum and come into contact with the *Purkinje system,* whose fibres spread to all parts of the ventricular myocardium. Bundle of His is normally the only conducting tissue between the atria and the ventricles.

   This is the fastest conducting system in the heart. Conduction through the Purkinje fibres is 1.5 to 4.0m/sec as compared to 0.4m/sec through contractile myocardial fibres.

The conducting system is composed of modified cardiac muscle. The SA node and to a lesser extent the AV node also contain small round cells with few organelles, which are connected by gap functions. These are probably the actual pace maker cells, and therefore called P *cells.*

Heartbeat and its Application

The rhythmic contraction and relaxation of the cardiac muscles is known as the *heartbeat.* Each cycle of the *heartbeat* consists of the following two phases:

1. the contraction phase termed *systole* (atrial systole followed by ventricular systole).
2. the relaxation phase called *diastole.*

Thus, a complete heartbeat consists of one systole and one diastole, the whole beat lasting about 0.8 sec. in a normal human adult at rest. Normally the heart beats about 72 times per minute and pumps out about 60 to 110 ml of blood per heartbeat.

The origin and spread of cardiac excitation and the electrical activity of the heart may be considered under the following headings:

(i) *Origin of the Heartbeat:* The heartbeat originates in the sinoatrial node (S-A node) which, therefore, has been called the *pacemaker* of the heart. This node is situated at the junction of the superior vena cava and the right auricle.

(ii) *Spread of the Impulse:* From this node the impulse spreads over the auricles causing their contraction and then, it stimulates the A-V node. This in turn stimulates the rest of the conduction system causing the rest of the heart to contract. The auricular phase of the beat lasts about 0.11 sec. Then auricles relax for the remaining 0.69 sec. of the cycle and thereafter they become ready for the next beat.

With the contraction of the auricles and during their relaxations period, the ventricles are fully distended and contract in their turn. Their thick walls generate much more pressure than the walls of the auricles. A more rapid ventricular contraction rate is made possible by *Bundle of His* and purkinje fibre system which is specialised myocardium for conducting impulses through the heart. This produces simultaneous contraction of ventricles to insure efficient pumping of blood through the *aorta* and *pulmonary artery,* because the A-V valves are already closed and prevent backflow of the blood into the auricles. The ventricular systole (contraction) lasts about 0.33 sec and they relax for about 0.47 sec (ventricular diastole). As soon as the ventricular phase of the heartbeat is over, the auricular phase starts again.

(iii) *Electrical Events in the Origin of Heartbeat:* The answer of how actually the wave of excitation or the impulse originates in the pacemaker can be satisfactorily given by considering the electrical events responsible for it. Interestingly, the sodium, potassium and calcium ions play an important physiological role in generating an electrical impulse in the cells of the sino-auricular node. Therefore, the balance between these ions as well as their absolute concentration is important.

The pacemaker tissue is characterised by an unstable membrane potential that slowly decreases after each impulse until the firing level is reached and another impulse is generated.

Ordinarily (at rest) the inside of the pacemaker cell is electrically negative with respect to the outside which is positive. When this condition exists the cell membrane is said to be *polarised.* This is due to the predominance of Na^+ ions outside the cell which outnumber the K^+ ions inside the cell.

At this time the potential difference is about -55 mV, and each cell thus becomes a small charged battery. However, the plasma membrane of pacemaker cells is partly permeable to Na^+ ions so that these ions continuously rush in the cells with a resultant decrease in the potential difference between inside and outside of the cells.

When the potential difference is reduced to about -40 tc -45 mV, the Na^+ ions rush in quickly while some of the K^+ ions escape outside. This leads to a reversal of electrical charges so that the inside of the membrane of pacemaker cell now becomes positive with respect to the outside. It is generally believed that this initial depolarisation represents the *pacemaker potential or pre-potential.* Thus, the initial depolarisation is due to Na^+ influx through rapidly opening Na^+ channels (the so-called Na^+ current, I_{Na})

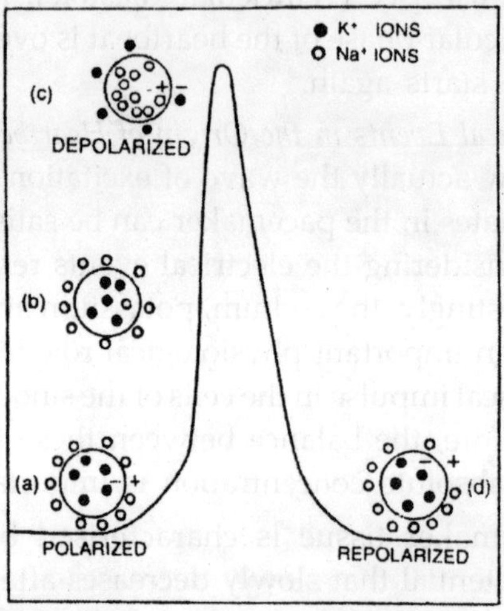

Electrical events in the origin of heartbeat.

When a certain level of depolarisation is reached by the pacemaker potential, a spike is generated. In other words, depolarisation causes the *action potential* and the electrical impulse generated spreads to adjacent cells by a kind of relay system so that the intensity of impulse does not decrease with the distance travelled. Depolarisation initiated in the SA node spreads radially through the atria and then converges on the AV node. Atrial depolarisation is complete in about 0.1 second. Because conduction in the AV node is slow, there is a delay of about 0.1 second *(AV nodal delay)* before excitation spreads to ventricles. From the top of the septum, the wave of depolarisation spreads in the rapidly conducting purkinje fibres to all parts of the ventricles in 0.08-0.1 second.

The depolarisation of the ventricular muscle begins at the

left side of the interventricular septum and moves first to the right across the midportion of the septum. The wave of depolarisation then spreads down the interventricular septum to the apex of the heart. It returns along the ventricular walls to the AV groove, proceeding from the endocardial to the epicardial surface. The parts of the heart to be depolarised in the last are the posterobasal portion of the left ventricle, the pulmonary conus and the uppermost portion of the septum.

At the peak of each impulse I_k begins and brings about *repolarisation.* I_k then declines, and as K^+ efflux decreases, the membrane begins to depolarise, forming the first part of the *pre-potential.* Ca^{2+} channels then open. These are of two subtypes in the heart, the *T channels* (T for transient) and the *L channels* (L for long-lasting). The calcium current due to opening of T channels ($I_{ca}T$) completes the pre-potential, and the calcium current due to opening of L channels ($I_{ca}L$) produces the impulse. It should be noted that resting membrane potential of pacemaker tissue is somewhat lower than that of atrial and ventricular muscle.

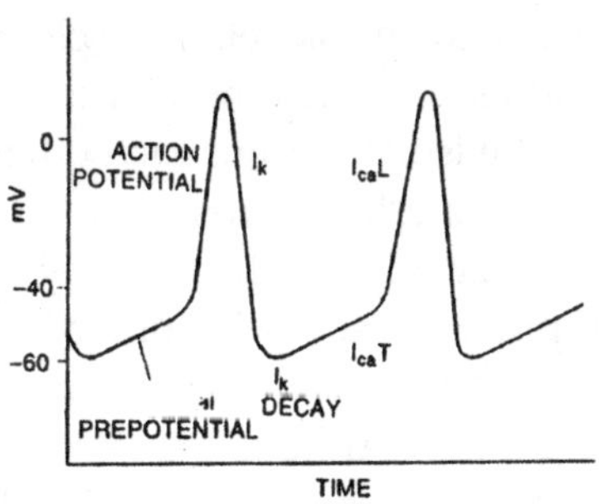

Diagram showing membrane potential of pacemaker tissue.

The depolarised cells soon repolarise by becoming negative inside and positive outside the plasma membrane. This is achieved by pumping out the Na^+ ions that had infiltrated earlier. Since the Na^+ concentration is already higher outside the cells, they have to be actively transported with the expenditure of the energy of ATP. The characteristic feature of the cardiac action potential is the long duration of the repolarisation phase. The rate at which the electrical impulses are discharged depends on several factors: The vagus nerve tends to reduce it while the sympathetic nerves tend to increase it. The other factors which can bring a change in the stability of the membranes and thus alter the rate of cell discharge include the pH, the O_2 and CO_2 concentration of the blood, hormones and drugs, etc.

All-or-none Principle of the Heartbeat: In skeletal muscles the extent of contraction (within limits) is proportional to the strength of the stimulus. In the heart this is quite different. The heart, as a whole follows the so-called *all-or-none-principle, i.e., if the heart muscle beats at all, it does so to its maximum effect.* Thus, the strength of the contraction is independent of the strength of the stimulus and once the impulse has been initiated by the pacemaker cells the high conductivity of the cardiac muscle cells transmits it rapidly.

However, this principle holds true also for the *individual fibres* of the skeletal muscle, although the muscle as a whole is capable of graded contractions. The reason for this difference between the heart and the skeletal muscle lies in the relation of the muscle fibres to each other. In a skeletal muscle the individual fibres are isolated from each other by a *sarcolemma,* but in the heart the imperfectly cross-striated fibres are brought into intimate contact with each other by anastomosis. This permits the impulse to spread from one fibre to another throughout the whole atrium or ventricle, causing them to respond as a single unit. Although the contraction of the heart is always maximal, this does not mean that the extent of contractions are always the same. They are determined by the physiological state of the heart muscle.

Starling's Rule

The length-tension relationship in cardiac muscle is similar to that in skeletal muscle. As the muscle is stretched, the developed tension increases to a maximum and then declines as stretch becomes more extreme. Starling pointed this out in the form of a law and noted that the *"energy of contraction is proportional to the initial length of the cardiac muscle fibre"*. This is also called Frank-Starling law or mechanism.

When a large volume of venous blood arrives at the heart, the increased diastolic filling stretches the fibres of the right ventricle. Up to a certain point, this stretching increases the force of the contraction, enabling the ventricle to discharge a greater volume of blood. Thus, according to the Starling's law of the heart, within physiologic limits, the more the ventricles are filled with the incoming blood the greater is the force of their contraction at systole. Its physiological significance is that as blood is returned more rapidly to the heart during exercise, the increased volume stretches the walls of the ventricles and causes a more forceful propulsion of blood out of the heart. This is one of the main factors balancing the amount of blood entering and leaving the heart.

Starling's law of the heart also operates in the opposite condition. For example, in extensive haemorrhage the systolic and diastolic pressures are very low. The amount of blood delivered to the ventricle is greatly reduced, and, as a result the cardiac fibres are only slightly stretched. This results in a feeble heartbeat and in a much lower than normal systolic ventricular pressure. However, this force is sufficient to overcome the much reduced diastolic pressure in the aorta.

Electrocardiogram (ECG) and its Characteristics

A heart beat can also be analysed in terms of electrocardiogram (ECG) obtained by electrocardiograph. Most electrocardiograph machines record the cardiac activity on a moving strip of graph paper. Typically the normal ECG consists of a series of slow waves, including upward (negative) deflections called P, R and downward positive deflections Q and S. Of them, (i) the P wave represents *atrial depolarisation,* which precedes the contraction of the atria. The PQ interval represents the delay at the auriculo-ventricular junction, (ii) The QRS complex represents *ventricular depolarisation* preceding ventricular contraction. A normal QRS complex lasts about 0.08 sec. (iii) The T wave represents *ventricular repolarisation*. It is much smaller than the QRS depolarisation complex. There is no comparative atrial repolarisation visible, for it is obscured by the large QRS depolarisation of the ventricles, occurring at that time.

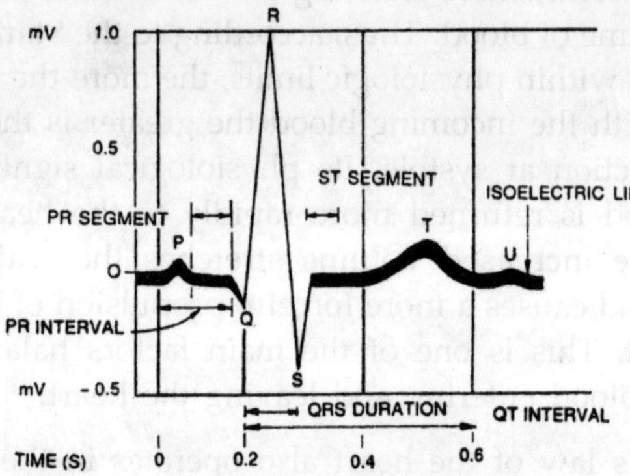

The waves of ECG.

An important parameter of the ECG is the *PR interval*, which indicates the length of time between the beginning of the atrial contraction and the beginning of the ventricular contraction, and thus the conduction time through the heart. A normal PR interval is 0.18 sec or less. In incomplete heart block, when conduction through the AV bundle is slowed, the PR interval may be almost doubled, and there may bè 2 to 3 P waves for each QRS-T complex. In complete heart block, the ventricles "escape" from SA and atrial control, and the P waves may be completely dissociated from the QRS-T complexes. Prolongation of the QRS complex, which normally lasts about 0.06 sec, indicates delayed conduction through the ventricles, often caused by ventricular hypertrophy. The QT interval coincides with the beginning and the end of ventricular contraction. It lasts about 0.40 sec. The electrocardiogram reflects the activity of various parts of the heart and thus aids in diagnosis of different heart ailments.

Sounds of Heart

Two sounds are normally heard through a stethoscope.

The first sound, usually heard as the "lub" of the "lub dup, lub dup" of the heart, is very low pitched, and it is caused by the closure of the mitral and tricupsid valves before the beginning of the ventricular contraction. The first sound has a frequency of 30-80/sec and a duration of 0.05 sec. The *second sound,* "dup", is higher in pitch. It has a frequency of 150-200/sec and a shorter duration (0.025 sec). It is caused by the vibrations associated with closure of the aortic and pulmonary valves just after the end of ventricular systole.

The third and fourth sounds are less important and can be detected on the phonocardiogram. The *third sound* is due to vibrations of the cardiac walls produced by the rapid filling of the ventricles. The *fourth sound* occurs during atrial systole. It has low frequency and amplitude.

The characteristics of the sounds are clinically important. The cause of *cardiac murmurs* is disease or disorder of the heart valves.

Cardiac Arrhythmias and Heart Block: In the normal heart each beat originates in the SA node. This is called normal sinus rhythm (NSR). The heart beats about 70 times per minute at rest. The rate is slowed *(bradycardia)* during sleep and accelerated *(tachycardia)* by exercise, fever, emotion and many other factors.

Heart block is a condition in which the normal cardiac rhythm is disturbed *(cardiac arrhythmias)*, or there is defective conduction in some part of the heart. This might occur as *sinoatrial nodal block, atrioventricular nodal block, infranodal block* or *right* or *left bundle branch block.*

Sinus node dysfunction leads to marked bradycardia

accompanied by dizziness and *syncope* (sick sinus syndrome). It is treated by implantation of pacemaker. When conduction between atria and ventricles is slowed but not completely interrupted, the condition is partial or *incomplete heart block.* In the form called, *first-degree heart block,* all the atrial impulses reach the ventricles but PR interval is very long. In the form called *second degree heart block,* not all atrial impulses are conducted to the ventricles. In another form of incomplete heart block, there are repeated sequences of beats in which the PR interval is progressively increased until a ventricular beat is dropped *(Wenckebach phenomenon).*

When conduction from the atria to the ventricles is completely interrupted, *complete* (third-degree) *heart block* results. The block may be due to disease in the AV node *(AVnodal block)* or in the corducting system below the node *(Infranodal block).* In case of AV nodal block the remaining nodal tissue becomes the pacemaker and the rate of the idioventricular rhythm (ventricles beat at a low rate) is about 45 beats per minute. In case of infranodal block due to disease in the bundle of His, the ventricular rate is 35 beats per minute or still lower.

If complete heart block develops suddenly, there may be a delay before the ventricles start beating at their own rate. Naturally, in this period the systemic blood pressure falls to a very low level and the blood supply to the brain is highly insufficient. It the ventricular standstill lasts for only a few seconds it may give rise merely to faintness and dizziness, but if it is more prolonged it may cause loss of consciousness, convulsions, and even death. An attack of complete heart block together with fainting constitutes the *Stork-Adams Syndrome.*

Role of Cardiac Output

The cardiac output is the amount of blood pumped by the left ventricle into the aorta per minute. Because this is the volume of blood returning to the heart from the lungs, it is also the same volume that was pumped to the lungs by the right ventricle. As a result, in the normal heart, the right and left sides of the heart pump the same volume of blood. The normal cardiac output for the young adult male, reclining is about 5.6 litres/min, that for a female of the same age and size is 5 litres /min. Cardiac output is affected by changes in posture and by exercise.

The relationship between *cardiac output, heart rate* and *storke volume* is expressed as follows:

Cardiac output = heart rate × stroke volume.

If the heart rate is 72 beats/min and the volume of blood pumped out by each ventricle per beat (stroke volume) is 70 ml, then the cardiac output at rest is 5040 ml/min. There is an integrated control of cardiac output. Regulation of cardiac output as a result of changes in the cardiac muscle fibre length is called *heterometric regulation.* Regulation due to changes in contractility independent of length is called *homometric regulation.*

Stroke volume depends largely on the volume of venous blood returning to the heart (Starling's law of the heart), the size of the heart, and its contractile strength. Trained athletes usually have hearts larger than normal individuals, because of hypertrophy of cardiac muscle fibres as a result of constant work against resistance. These muscle fibres have more contractile proteins and myoglobin than is found in untrained individuals, and consequently they contract with greater strength. This increase in the inotropic force of the heart increases the stroke volume and permits the athlete to achieve the same cardiac output with a slower heart rate. This is not the case in pathological enlargement of the heart, with which the athlete's heart should not be confused. An interesting problem is the mechanism by which the athlete's heart rate is slowed by training. However, all individuals do not respond to the same extent, and it must be associated with a felicitous combination between the vagal inhibitory nerves and the inherent heart rate. The athlete's heart rate may be as low as 45 to 50 beats/min, but the increased stroke volume of 100 to

110 ml/ min results in normal cardiac output, the advantage of this is that the athlete has to increase heart rate proportionally less during strenuous exercise than does the untrained individual. This permits better filling of the heart in diastole and contributes, together with the greater strength of the hypertophied heart, to an enhanced cardiac output..

Regulation of Cardiac Activity

(a) *Reflex Control (Nervous and Chemical Rcgulation):* The pacemaker is the immediate regulator of the heart rate. But the action of the pacemaker is controlled by the *heart rate centre* located in the *medulla oblongata.* Two pairs of nerves lead from the heart rate centre to the heart. Impulses through one pair, formed by a sensory and a motor nerve, accelerate the heart beat, and impulses through the other pair also formed by a sensory and a motor nerve, slow the heart beat. The inhibitory nerve fibres (parasympathetic nerve fibres) pass from the heart rate centre into the large vagus nerves. Both the accelerator and inhibitor nerves terminate in the right auricle above the pacemaker. Sensory nerves lead to the heart rate centre from the walls of the vena cava and the aorta. When the pacemaker is stimulated, a wave of contraction spreads out from it through both auricles and ventricles. The vagus nerve fibres inhibit the heart rate by releasing *acetylcholine* (Ach). The accelerator nerves (sympathetic nerve fibres) increase the heart rate by releasing *sympathin.* The underlying mechanism is that the acetylcholine, which is released at the synapse (the small gap between the vagus nerve fibres and the pacemaker cell), increases the permeability of the cell membrane to the K^+ ions so that they pass into the cell increasing their negativity (hyperpolarisation). Under such conditions, the time between two successive depolarisations would be longer and consequently the rate of heart beat slower.

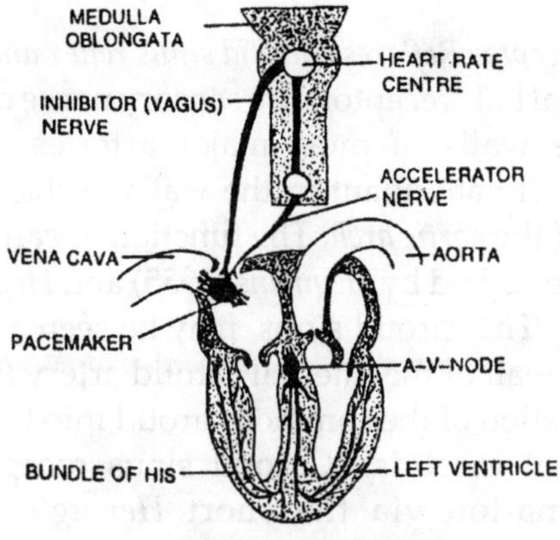

The motor innervation of the heart.

The firing rate of these two groups of motor fibres (sympathetic or cardioacceleratory, and parasympathetic or cardioinhibitory) is controlled by the antagonistic centres located in the medulla, the cardioaccelerator and cardioinhibitory centres. The purpose of this control system is to increase heart rate whenever blood volume or blood pressure falls in the main arterial system and to decrease heart rate whenever blood volume or pressure increases above a normal level.

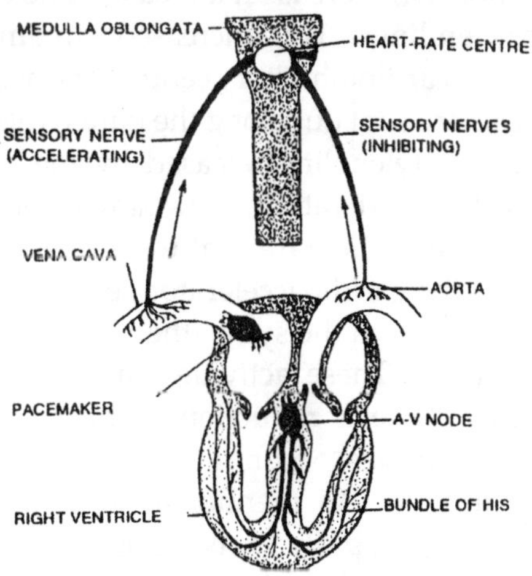

The sensory innervation of the heart.

(b) *Baroreceptor Reflexes (carotid sinus reflex and aortic reflex)*: Arterial baroreceptors are sensory nerve endings found in the walls of most major arteries, but they are especially abundant in the walls of the *carotid sinuses* and of the *aortic arch.* The function of carotid sinus was first described by *Heymans* (1955) and *Heymans and Neil* (1958). The carotid sinus may be seen as a thickening in the wall of the internal carotid artery (just above the bifurcation of the common carotid into the internal and external carotids). Carotid sinus receptors transmit information via the short *Hering's nerve* to the glossopharyngeal nerve (IX cranial) and then to the medulla. *Aortic receptors* transmit information to the medulla via the vagus nerve (X craninal). These baroreceptors are most sensitive within the normal physiological range of arterial pressure. Below about 60 mm Hg pressure the receptors do not fire at all and maximum rate of firing is attained at about 160 to 200 mm Hg. Thus their rate of firing is increased proportionally with increased stretch on the arterial walls

When the pressure in the aorta or carotid sinus goes above normal, there is an increased rate of firing of the sensory endings. This increased stream of impulses causes the cardioinhibitory centre to be activated. Nerve impulses are sent out along the parasympathetic vagus and the acetylcholine released by the nerve endings causes the heart rate to slow down. Simultaneously, inhibitory impulses are sent from the cardioinhibitory centre to the caridioaccelerator centre, and the stream of impulses from the sympathetic neurons to the heart is decreased. These activities thus slow down heart rate and cause a reduction in blood pressure. The reverse activities occur if the blood pressure drops below normal. As mentioned earlier baroreceptors respond most rapidly to changing blood pressures, the rate increases even during systole and decreases during diastole of the heart.

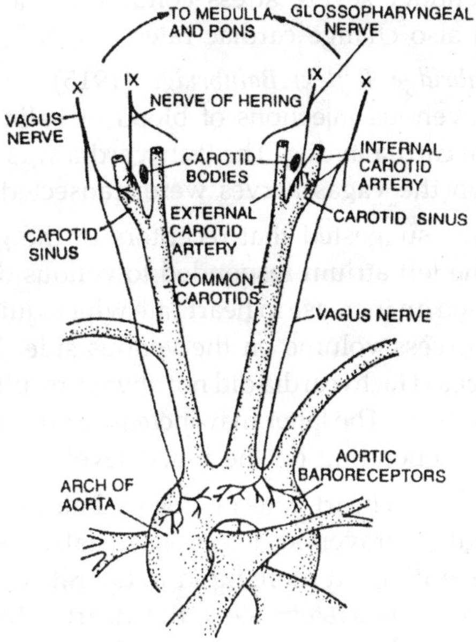

The arterial chemoreceptors; the aortic and carotid bodies.

The baroreceptors have sufficient sensitivity to regulate and maintain blood pressure constant during changes in body posture, as when a human sits down or stands up. It may also be noted that strong pressure exerted upon the carotid sinuses can result in unconsciousness due to the high pressures causing the heart to slow down or even stop.

In the latter case, the heart usually begins to beat again after about 10 seconds due to the phenomenon of *vagus escape*—the vertebrate heart upon continued vagal stimulation is usually found to take up its own inherent rhythm due to the pacemaker and ignores the vagus excitation.

Chemoreceptor endings are found in the aortic arch and in the carotid arteries. These are primarily concerned with controlling oxygen and CO_2 levels in the blood by their actions on the respiratory system.

But under severe stress conditions their stimulation will also change cardiac rates.

(c) *Bainbridge Reflex: Bainbridge* (1915) discovered that intravenous injections of blood or saline produced a tachycardia in dog. The tachycardia was not produced when the vagas nerves were transected.

It was suggested that receptors in the great veins or in the left atrium responded to venous distention and caused an increase in heart rate which quickly removed the excess volume on the venous side. However, the expected tachycardia did not always result upon venous distention. The latter may increase or decrease the heart rate, depending on the initial level of the heart rate.

When the heart rate of dog is below 140 beats per minute, intravenous injections usually result in cardiac acceleration. At more rapid initial rates, injections into the venous system slow the heart. Thus, the reflex seems to be a mechanism for keeping the heart rate within a given range. However, the nervous pathways of this complex reflex are at present not known.

(d) *Integration of Reflex Control*: There are connections between the cardioaccelerator and cardioinhibitory centres of the medulla and also connections between these centres and other regions of the brain, hypothalamus and cerebral cortex.

The heart rate can be altered by activity in higher centres. It is well known that various emotional states can accelerate or slow down the heart rate, the reason being that states can accelerate or slow down the heart rate, the reason being that emotional centres in the cortex and centres in the hypothalamus provide inputs into the medullary centres.

Thus, suprabulbar centres have been shown to be of utmost importance in the integration of reflex control of the cardiovascular system.

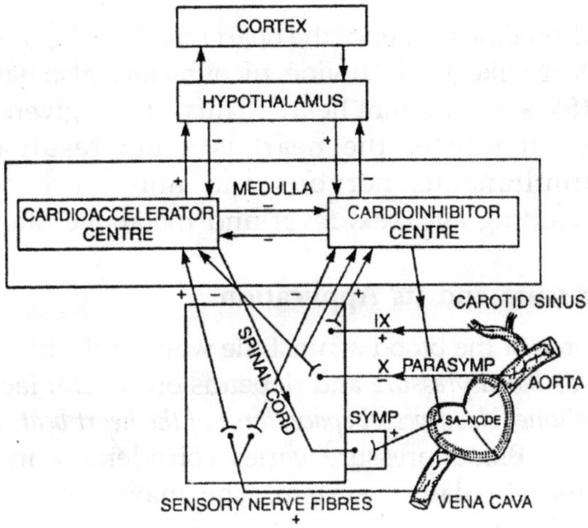

Integration of reflex control of cardiac activity.

(e) *Endocrine Control:* Endocrine glands play an important role in the control of heart rate and stroke volume. The release of adrenaline by the adrenal medulla is under the partial control of sympathetic nerves influenced by the cardioaccelerator centre in the medulla. In the flight or fight reflex of mammals, adrenaline is released into the bloodstream and among other actions increases the heart rate and also the contractile force of the cardiac muscle. The adrenal medulla and other endocrine glands also aid in maintaining the proper metabolic and ionic environment for correct activity of the heart.

(f) *Other Factors Regulating Cardiac Activity:* Besides the nervous control described above other factors also regulate the heartbeat. Most of them act directly on the heart muscle or on the pacemaker. Like any other metabolic process, the heartbeat is affected by temperature. High and low temperature accelerate and deaccelerate the heartbeat respectively. An acidic pH accelerates the heart, while the alkaline pH retards it. A high CO_2 concentration in the blood also increases the heart rate since it lowers the pH of the blood. The mineral ions in the blood and hormones like thyroxin, insulin, sex hormones, pituitary hormones, and adrenaline influence the heart rate directly or indirectly. Drugs like acetylcholine, nicotine etc., also have varied effects on different hearts. Thus, at any given moment, actual rate of the heart is a net result of many simultaneous nervous and non-nervous effects, including age, sex, fever and metabolic rates.

Blood Pressure and its Application

The force of the blood against the walls of the blood vessels is known as *blood pressure* and depends on several factors such as *blood volume, blood vessel space, force of the heart beat,* and *blood viscosity,* etc. Blood pressure varies considerably in different parts of the circulatory system. The maximum pressure is encountered in the *aorta* where it normally reaches to 140 mm of mercury. Then it gradually decreases in the arteries and arterioles, becoming lower in the capillaries and still lower in the veins. The arterial blood pressure undergoes a rhythmic change reaching a maximum during ventricular systole when blood is pumped into the arteries and decreasing during ventricular diastole. The former is called *systolic pressure* and the latter, the *diastolic pressure.* The normal systolic pressure in humans ranges from about 120 to 140 mm of mercury, whereas the diastolic pressure ranges from about 75 to 90 mm of mercury.

Factors of Blood Pressure

Blood pressure is affected conciderably by numerous factors including age, sex, heredity, and other physical and emotional states. However, there is controversy about where to draw the line between normal and elevated blood pressure levels (hypertension), particularly in older persons. Because of nervousness, about 20 per cent of hypertensive individuals have high blood pressure in the doctor's room than during their normal daytime. This is called white coat hypertension.

An important cause of the systolic pressure rise with age is decreased distensibility of the arteries. At the same level of cardiac output, the systolic pressure is higher in old persons than in young ones because there is less increase in the volume of the arteries during systole to accommodate the same amount of blood.

The *total blood volume,* affecting the blood pressure, is adjusted to some extent through the contractions of the spleen which bring stored blood into circulation. But adjustments of blood pressures by such means are comparatively slow. However, more effective adjustments are brought about by alterations in the force of the heart beat and the space within blood vessels. The force of the heart beat is determined by the *amount* of blood received and pumped out by the heart.

Operation of Blood Pressure

Vasomotor Control: The space within the blood vessels is adjusted by contractions and relaxation of the muscles in their walls. These muscular activities are controlled by a *Vasomotor centre* in the brain, located in the medulla oblongata.

The vasomotor centre consists of an *excitatory portion* that continuously fires sympathetic vasoconstrictor nerves to maintain vasomotor tone and an *inhibitory portion* that can inhibit sympathetic vasoconstriction, thereby allowing passive vasodilation. Nerves lead from the vasomotor centre to all blood vessels except the capillaries which do not have muscles. *Vasoconstriction* reduces the diameter of blood vessels, *vasodilation* increases it. Vasoconstriction and vasodilation together are referred to as *vasomotion.*

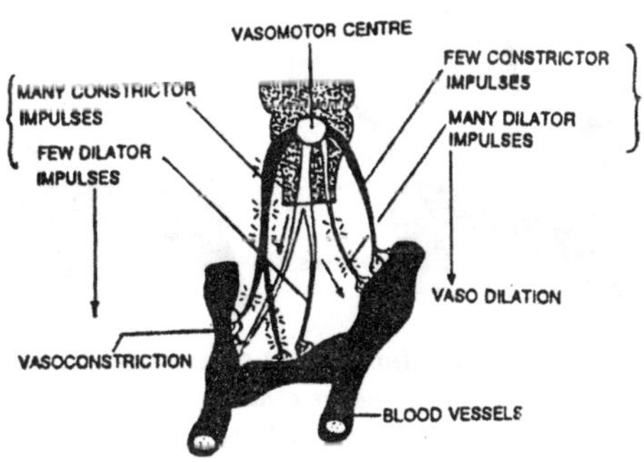

The pattern of vasomotion.

The pattern of vasomotion indicates that vasoconstriction occurs when a blood vessel receives many constrictor impulses and a few dilator impulses from the vasomotor centre. Vasodilation occurs when dilator impulses are many and constrictor impulses a few. Vasoconstriction and localised vasodilation follow through the numbered events shown,

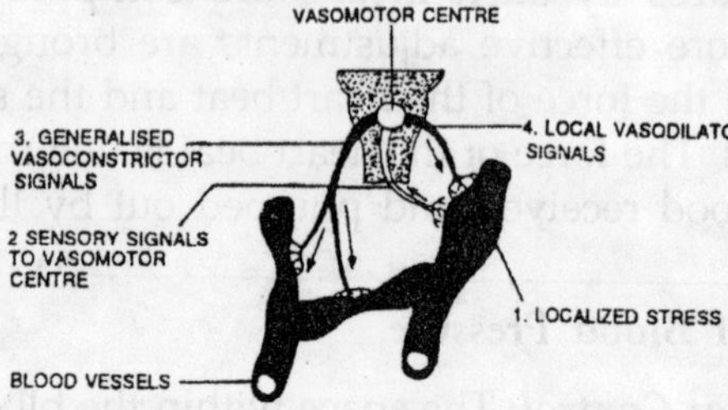

The vasomotor effects of localised stress. Generalised vasoconstriction.

raises the overall blood pressure, vasodilation lowers it. Pain, emotions and stresses generally all tend to increase the blood pressure through bodywide vasoconstriction. At the same time, in specific parts of the body which sustain a stress (like a wound, etc.), the local blood pressure often falls through local vasodilation resulting in swelling. The adaptive advantage of this is clear. A rise in the overall blood pressure enables the animal to respond to the stress more effectively, while local vasodilation itself permits more blood to flow into that region, thus making available more nutrients and oxygen.

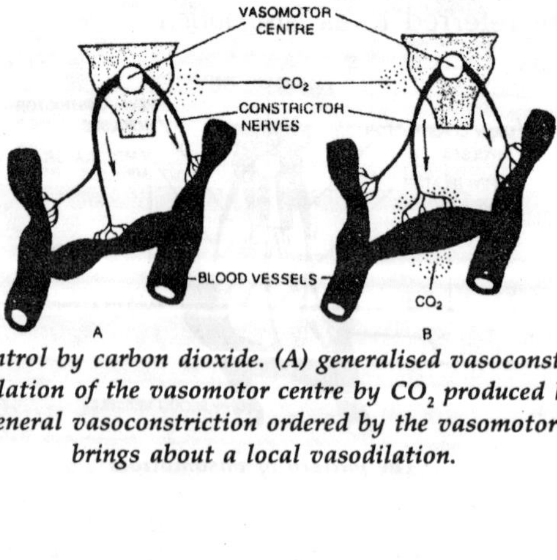

Vasomotor control by carbon dioxide. (A) generalised vasoconstriction as a result of stimulation of the vasomotor centre by CO_2 produced locally over-rides of the general vasoconstriction ordered by the vasomotor centre and brings about a local vasodilation.

Like the heart rate the blood pressure is normally prevented from varying excessively by stretch reflexes. Some sensory nerve fibres from the aorta lead into the vasomotor centre. When the aorta is greatly distended by blood spurting out from the heart, *i.e.*, when blood pressure is high, a stretch reflex via the vasomotion centre brings about body-wide vasodilation. In this manner a high blood pressure is reduced to normal. Conversely, when the aorta distends very little or not at all, a stretch reflex leads to increased vasoconstriction and a rise of blood pressure.

The carbon dioxide of the blood also affects the vasomotor centre. If the blood reaching the vasomotor centre carries a high concentration of CO_2, then the centre transmits constrictor impulses throughout the body and the blood pressure will rise. Conversely, low CO_2 concentrations result in a fall of blood pressure.

The blood pressure is adjusted to by variations in the rate of tonic discharges of sympathetic nerves that constrict arterioles and veins and increase heart rate and stroke volume. This is called feedback control of blood pressure.

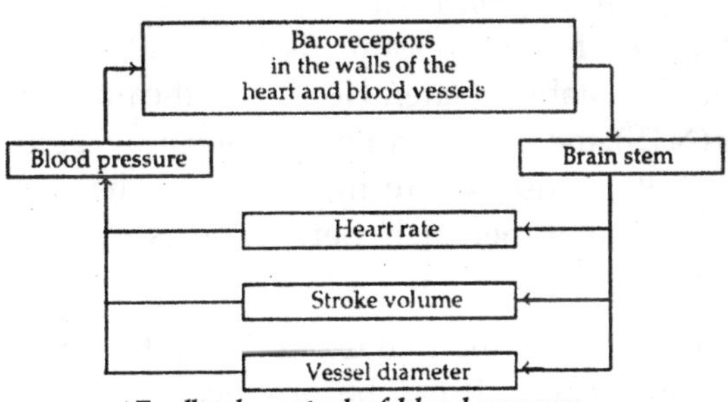

Feedback control of blood pressure.

Significance of Hypertension

The blood pressure in the aorta and in the brachial and other large arteries in an adult human rises to a peak value *(systolic pressure)* of about 120 mm Hg during each cardiac cycle and falls to a minimum value *(diastolic pressure)* of about 70 mm Hg. A sustained elevation of the systemic arterial pressure is called *hypertension*. In some cases *Pulmonary hypertension* can *also* occur, but the pressure in the pulmonary artery is relatively independent of that in the systemic arteries.

Some of the procedures that cause *experimental hypertension* in certain animals involve manipulation of the kidneys, nervous system, or adrenals. Some strains of rats develop hypertension quite spontaneously (SHR rats), or when given a high sodium diet (Dahl strain rat)

The hypertension that follows constriction of the renal arterial blood supply or compression of the kidney is called *renal or Goldblatt hypertension*. This may or may not be dependent on plasma renin activity. The decreased ability of the constricted kidney to excrete Na^+ is an additional factor contributing to renal hypertension.

Bilateral clamping of the carotid arteries proximal to carotid sinuses elevates the blood pressure and heart rate because the procedure lowers the pressure in the sinuses. If carotid sinus nerves are cut on each side, the same effect is obtained. The pressure rise following these two procedures is moderate as the aortic baroreceptors are still functioning normally. If baroreceptor afferents in the vagus nerves are also interrupted, the blood pressure rises to 300/200 mm Hg or higher and is unstable. Bilateral lesions of the nucleus of tractus solitarius (NTS), the site of termination of baroreceptor afferents in the medulla, cause severe hypertension that can be fatal. This form of experimental hypertension is called *neurogenic hypertension*.

In humans, sustained diastolic hypertension is a very common abnormality. It can be produced by many diseases including adrenocortical diseases, tumours of adrenal medulla or of catecholamine-secreting tissue elsewhere in the body, renal diseases, narrowing of the aorta, severe polycy-themia and by oral contraceptives ("pill hypertension"). Hypertension is also a prominent symptom of *toxemia of pregnancy*, which may be caused by a pressor polypeptide secreted by the placenta. Hypertension is also seen in Cushing's syndrome.

In most of the humans with high blood pressure, the cause of the hypertension is unknown and they are said to suffer from *essential hypertension.* In the early stage of hypertension, the blood pressure elevations are intermittent, but later the blood pressure elevation is continued. The baroreceptor mechanism is "reset" so that the blood pressure is maintained at the elevated level. In women, untreated essential hypertension is often benign, but it may also progress rapidly into malignant phase.

Chronic hypertension can enter an accelerated phase in which necrotic arteriolar lesions develop and there is a rapid downhill course with papilledema, cerebral symptoms and progressive renal failure. This syndrome is called *malignant hypertension* and is very fatal if not treated. However, its progression can be checked and it can be reversed by appropriate anti-hypertensive therapy.

Hypertension causes a number of serious disorders including hypertrophy of cardiac muscle or heart enlargement. There is an increased incidence of atherosclerosis in hypertension and myocardial infarcts are common even when the heart is not enlarged. Starling's law operates in hypertension, but there may be heart failure. Hypertensive individuals are also predisposed to thrombosis of cerebral vessels and cerebral hemorrhage. An additional disorder is renal failure. However, the incidence of heart failure, strokes, and renal failure can be considerably reduced by active treatment of hypertension, even when the hypertension is relatively mild.

At present, essential hypertension is a treatable but not a curable disease. Drugs that block α-adrenergic receptors, β-adrenergic receptors, and drugs that inhibit the activity of angiotension - converting enzyme, and calcium channel blockers that relax vascular smooth muscle can produce effective lowering of the blood pressure. In humans, several of the diseases known to cause hypertension are not only treatable but also curable. When fed on high sodium diet some humans show marked increase in blood pressure. As there is no easy test for salt-sensitivity (to distinguish salt-responsive from salt-resistant humans), the current emphasis on less sodium intake in all humans appears to be beneficial.

Significance of Hypotension

The term hypotension describes the condition or symptoms produced by sustained low blood pressure. The low pressure receptors are those in the great veins, right and left atria and pulmonary vessels. Hypotension is also a common abnormality in humans. It can be produced by several disorders and diseases. Hypotension and hypovolemia produced by hemorrhage release large amounts of vasopressin resulting in water retention and reduced plasma osomolality. There is exponential increase in plasma vasopressin produced by decreases in blood pressure.

In cases of *Orthostatic (posturnal) hypotension,* which is common in patients using sympatholytic drugs, sudden standing causes a fall in blood pressure, dizziness, dimness of vision and even fainting which is called postural syncope. Postural hypotension may also occur in diseases such as diabetes and syphilis in which there is damage to the sympathetic nervous system. Another cause of postural hypotension is *primary autonomic insufficiency.* It may be in several forms including one in which there appears to be a cogenital deficiency of dopamine β -hydroxylase and consequently no production of epinephrine and norepinephrine.

Fainting or syncope, more commonly due to peripheral, vascular or cardiac abnormalities causing inadequate cerebral blood flow, produces hypotension together with bradycardia (slow heart rate). This condition is called *vasovagal syncope.* Patients with orthostatic (postural) hypotension also show *micturition syncope,* fainting during urination. It is due to combination of orthostasis and reflex bradycardia induced by voiding in these patients.

14

Biological Clocks

You must have all experienced that you tend to wake up almost at the same time everyday in the morning and you start feeling sleepy at certain time as the night falls or why do we so often get up before the alarm is due to sound? Because there is a built in clock inside your body, which regulates sleep-wake timings. It has now been established that all organisms possess an innate biological clock (Pittern drigh, 1960).

Rhythms of various kinds have been in evidence in the biological world for quite sometime. It was during Great Alexander's march to India that one of his officers Androsthenes noted a daily leaf movement among the plants of Tamarind. About 300 years before Aristotle, men first wrote about plants which raise their leaves to the sky during the day and fold at night.

Charles Darwin noted rhythms in many animals and wrote in 1880 that the periodicity among plants and animals was to a certain extent inherited. In animals rhythms were observed less as compared to plants because an animal's daily routine is more complex than that of a plant. Rhythmic events occur every wherein the environment, the sun rises and sets forming days and nights, the moon moves through its phases, the tides get high and low regularly, constituting lunar and tidal rhythms respectively.

The orbit around the sun and rotation of earth creates seasons. All the living organisms are subjected to these rhythms, and they exert great influence on them. Animals must regulate their activities in concurrence with these cycles in order to avoid adverse conditions and to get enough food and shelter. Now it is Well known that daily rhythms form the basic feature of living organisms - from Protozoa to Primates. The Paramecium will congregate for reproduction only at certain times of the day, animals requiring warm temperatures, will become active during the middle of the day. The birds sing mostly in the mornings and evenings; owls, hyenas, jackals, and porcupines are active at night, whereas deers, monkeys and apes are active during the day time. People have always wondered on the sense of time exhibited by birds, bees, worms and plants in their day to day behaviour. A built in clock makes the owls, rats and cockroaches sleep during the day, bats come out of their roost to forage in the evenings, the whole day they remain upside down hanging by the branches; parrots, mynas, squirrels, house sparrows would sleep by night, the bean plants fold their leaves during the night and open in the day. Cold blooded animals remain active in the day time during winters and during mornings, evenings in summers; during extreme cold or heat they hide at appropriate places. You must have noticed that your pets specially dogs and cats lose more hair during summers, in winters their fur coat gets thicker. Domestic sheeps have an annual cycle of wool production, rodents, house lizards, squirrels exhibit cycles of hibernation and activity.

Their body weight is maximum when they go into hibernation because they have to depend on reserve food material and the weight is minimum when they wake up. Birds and other animals migrate at particular times in an year.

Behaviour exhibits a marked periodicity in its pattern in performance, and this periodicity involves complex interactions between nervous and endocrine systems and outside world.

In earlier times the most outstanding work on biological clocks was done by Erwin Bunning, in 1936 he proposed that organisms have an inbuilt and nearly automatic rhythm which repeats itself at about 24 hours interval and these rhythms are used to measure time.

Biological clocks are also known as internal clocks or circadian rhythms these rhythms are the internal mechanisms that provide a means of measuring length of time to the living beings. These rhythms allow the animals to adapt their behaviour in the best possible way to the course of a day's changing event. By setting the internal clock, the plants clock proves itself when its petals open at dawn and close at dusk. The common example around us is of a flower called as office flower or morning glory, or Portulacca which opens when the sun comes up and closes when sun sets. The birds leave for their migration at a pre-determined time to arrive at the destination at a suitable time. A dormouse cannot withstand extreme of cold weather, hence it goes into hibernation but comes out for food every three weeks, that too always near sunset. A human being's clock becomes evident after we fly to another continent, through changing time zones; we yawn through lunch, go to sleep when it is time for evening tea, wake up at two in the middle of the night and fall off to sleep at 8.00 am, but after few days the biological clock resets according to the new environment, the time taken in resetting the clock may vary from person to person.

It has been a subject of interest whether periodicity in an animal is determined by external environmental stimuli (exogenous) or controlled through physiological factors within the animal itself (endogenous). Both the stimuli are important. Their intimate relationship was brought to notice as early as 1960 by Lorenz, neither exogenous nor endogenous stimuli are solely responsible in controlling behaviour through biological clocks, it is the intricate interaction between the two together. Initially, to set the clock and run it, external stimuli like day length, lunar phases, tides, and temperature play the most important role but after some time the clock is maintained and run by internal stimuli controlled by body physiology. Experiments have shown that many animals maintained their biorhythms even when the relevant external stimuli were either removed artificially or when organisms were displaced from their habitat and placed in conditions of constant light intensity, temperature, pressure, humidity, and chemical composition, they continued to exhibit nearly the same rhythm of behaviour they showed in their natural habitat. In other words, animals possess internal clock that controls the rhythm in behaviour, and keep it going with the help of internal stimuli while external stimuli serve only to set and reset this clock periodically; such external environmental stimuli are called Zeitgebers or Entrianers because they entrain internal clocks and the process is called entrainment, an individual's clock can be re-entrained if placed in a new time zone, as happens with rapid, long distance trams-continental travel.

During evolution, animals have acquired a variety of endogenous rhythms, their periods are matched with rhythmic events in the environment. Some rhythms are matched to 24 hour cycle of light and dark (circadian), other to the 12.4 or 24.8 hour tide cycle (circatidal), the 29 day lunar cycle (circalunar) the yearly seasons (circa-annular) or the time between successive spring low waters (14.7 days; semi-lunar or circasyzygic). Prefixing word circa (which means about) with all kinds of biological rhythms is very necessary because all these internal clocks are only approximately matched, e.g. 24 hour cycle or rhythm does not mean 24 hours by wrist watch, it means about 24 hour or circadian.

Significance of Circannual Clocks

The activity of animals or plants, influenced by the seasons occurring once in an year remain under the control of circannual clocks. People who have been interested in gardening know that some flowers bloom only in summers while others like pansy, calendula, candituft and carnation bloom during winters because flowering in these plants is influenced by day length, temperature and humidity. Migration among birds and other animals to escape harsh climatic conditions and in search of food is a popular example. Migratory birds undertaking long distance migrations every year, show cyclic changes in their body weight, gonad size, plumage and colouration.

The larvae of many insects viz parasitic wasp, flesh fly show two phases every year (i) active (when the insect executes work actively) and (ii) inactive (diapause: a period of arrested growth and development in insects which is under the control of endocrine system), it is an adaptation to avoid adverse condition, but does not automatically end with the termination of the adverse conditions, it is genetically determined and can be broken by an appropriate environmental change or by a biological clock. Dormancy state in which warm blooded animals reduce their metabolic activities to a minimum level during unfavourable conditions so as to survive until conditions improve is also a cycle pattern, e.g. ruffed grouse do not migrate, instead, they bury themselves under the snow during the winter to lead a dormant life; several other birds, viz. small passerines also bury themselves in similar way. Humming birds move into caves from high Andes to escape below freezing temperature in winters.

This cyclic movement from Andes to caves and back remains under the control of circannual clock. Long term torpor a state of inactivity, similar to dormancy, into which some warm blooded animals (birds and mammals) enter in order to avoid adverse conditions and to conserve energy, it can be either long term or short term, is also influenced by circannual rhythms (Bartholomew, 1972). All the behaviours are tied to a clock so perfectly and the condition are so predictable that animals go in dormancy in anticipation. Hibernation and aestivation are also states of inactivity into which some cold blooded animals like frogs, toads, lizards and snakes enter in order to avoid cold (hibernation) and heat (aestivation). The timely escape from harsh weather conditions is achieved by circannual clock.

Monarch butterflies of northern states of USA and Canada show circa-annular rhythmicity in their annual migration to and from Mexico. From USA, monarchs migrate in millions every year in September covering about 3,200 km, they pass their winter in state of semi-hibernation on trees and in March they fly back. There is a small jungle near San Fransisco, known all over the world for its butterfly trees, because these trees remain covered by monarch butterflies from October to April. American thirteen lined ground squirrel generally begins to hibernate in October and emerges from its sleep during the following April.

This annual rhythm of hibernation could be seen even when external conditions were kept same for the squirrel in laboratory conditions. Almost everyone has witnessed that the common Indian house lizard disappears during winters. Fish, bird, and whale migration, movements of thousands of wildebeests every year in Africa and annual reproductive cycles in large vertebrates are some of the common examples of circannual rhythms. It is true that activities of many animals are synchronised with exogenous stimuli but what happens when animals are kept in unchanging conditions?

In an experiment millow warblers were subjected to constant twelve hour day for a period of fifteen months. These birds moulted twice in the year as they would have normally done and showed migratory restlessness at the usual time of the year (just like other free living' conspecifics). It must be noted that even under such constant experimental conditions, the periodicity was maintained by warblers for three years, after that it -got' disturbed and a readjusted itself according to the changed stimuli.

Significance of Circatidal Clocks

The biological rhythm synchronised with the low and high tides (the alternate rise and fall of the sea due to gravitational pull exerted by moon) in the sea are called circatidal clocks, the organisms living in the intertidal zone of the seashore are alternately submerged in water and exposed to air, due to this, pressure, salinity, food supply, temperature and predation risks change too, therefore, animals inhabiting tidal area show behavioural periodicity associated with the tides, e.g. some of the Planarians and Polychaetes which make burrows in shallow seawater or in tidal areas show vertical movement or movement out of their burrow as the high tide approaches; bivalve Molluscs increase filteration rates when high tide brings in more food; the beautiful column like Coelentrate, sea anemone increases the rate of body expansion and contraction during high tide, waving its tentacles actively to trap more and more food. It is noted that some animals get more active when the high tide approaches and some go deep into their burrows and will come out only after the tide has receded, it can lead to hide and seek between prey and predator, if the predators are coming out of the burrow it is beneficial for the possible prey to go deep down into the burrow. Significant work has been done by the marine biologists of California, on circatidal rhythms.

It was found out in one of the experiments, that sand beach hopper dig and make a burrow during the low tide and come up to swim at high tide, contrary to this, odd looking fiddler crabs leave their burrow when the tide has receded in search of getting food which the high tide must have brought and go back to the burrow when its time for high tide to flood the shore. At the same Californian coast, hundreds of small silver fish Grunion are brought to the coast alongwith a high tide. During their breeding season, the fish are washed on to the moist sand, where they spawn and then flip back towards water.

The time sense of laying eggs is so precise that their eggs hatch in fifteen days when the next high tide arrives, as the wave lashes against the sandy beach the baby fish are freed from the egg pods and start their journey into sea. A small crab, Synchellidium found in California inhabits the tidal zone, when the tide is low, these crabs bury themselves deep inside the wet sand, and during a high tide, when there is water, they come up to get food. Once this movement gets synchronised using exogenous stimuli, in this case, high tide, it was observed that crabs came out regularly in anticipation even in the absence of water which was stopped by putting barriers. There are air-breathing crabs which live on the seashores, they predict the arrival of high tide and start building a shelter in anticipation against the incoming water and predatory fishes, the soldier crab of tropical Asia starts forming a shallow crater, then pushes pellets of wet, muddy sand up on either side until it has built a roof over its head, the crab then burrows deeper, loosening more sand and plastering it overhead to form a sturdy roof, the shelter so built provides life giving inside, all those activities of this crab are intimately linked with tides and are controlled by circatidal clocks.

Functions of Circasynodic Clocks

Biological clocks which are synchronised with the moon phases are called circalunar or circasynodic clocks. The fantasised stories about warewolves have been known to many, that some persons change into a wolf and behave like animal on full moon nights, though this is truly a fiction, but in nature there are many animals specially marine invertebrates who transform into entirely different forms depending upon the phases of moon and perform particular behaviours, e.g. a marine Polychaete Platynereis dumerii is a long worm like creature, which lives in a burrow, on the, lateral sides of its body are flap 'like projections called parapodia, during their breeding season this seaworm becomes brilliantly colourful, gets bigger in size, its parapodia, eyes and tentacles also get enlarged, and it assumes a sexual form called heteronereis, which swarm actively at seashores during the full moon, perform nuptial dance and spawn. In the beginning platynereis and heteronereis were considered as two different species, because they looked so different latter it was realised that they are the non-sexual and sexual forms of one species.

The time bound activities of animals have been known from time immemorial. People of Fiji islands relish eating marine worms called palolo (Eunice viridis), which is a deep sea creature extremely difficult to catch but on the first day of the 1st quarter of the moon in November, the palolo worms swim to the ocean surface in large numbers, they swarm and spawn. Just before sunrise people get into their boats, over the coral reefs where the worms collect, it is a bonanza for people of Fiji who strain worms with large nets. Some terrestrial animals also show circalunar rhythms. Pit building activity in antlions (Myrmeleon obscurus) also reaches a peak around full moon.

Role of Circasyzygic Clocks

The biological rhythms which are synchronised with the fortnightly cycle of spring tide (high tide occurring a day or two after the new or the full moon) and neap tide (low tide which occurs in the middle of the second and fourth quarters of the moon) are known as semi-lunar clocks, e.g. pariwinkle, Littorina rudis shows a marked 14.5 day periodicity in its locomotory activity; the species lives high up on the shore and is only covered by the high water of spring tides when it comes out of its burrow to move around and feed. Aplysia, a mollusc also shows distinct 15 day periodicity in egg laying behaviour.

Role of Circadian Clocks

Operating on an approximately 24 hour basis, they are the most studied and best understood biological clocks and are found in almost all the major taxonomic groups of animal kingdom. In their simplest form they are reflected in the alternating periods of activity and sleep which correlates with the light/dark cycle.

In this cycle some animals remain most active at sunrise and sunset times, such animals are called crepuscular, there are animals who are active during the nights (nocturnal) but most animals are active during the day time, called diurnal. In an experiment, a flying squirrel was taught to turn a wheel in the day time and rest during the night, after that the animal was kept in continuous darkness. The observations showed the remarkable precision in the onset and cessation of wheel turning even in the complete absence of the environmental cues to which it was related.

Circadian or daily rhythm is best illustrated by the activity of bees. In an experiment bees were trained to visit a food dish at a certain time of the day, say between 12.00 to 2.00 in the afternoon when the sun shines to its maximum and temperature is also high; large number of bees learned to go to food dish between this time and they kept visiting the food dish roughly between this time even when there was an overcast, and the temperature was also lower; the bees worked according to an inbuilt clock.

Some intertidal crabs exhibit cyclic variation in their shell colour. They are lighter during the day time and get darker as the night falls, which prevents their detection from predators. The internal clock of bees did not correspond exactly to 24 hours, it was little shorter than 24 hours. Therefore, daily biological clocks are called circadian rhythms (i.e. approximate rhythms). Welly (1962) noted that snow buntings spend part of the day under snow during extreme cold.

Bats, rodents show daily torpor under extremely unfavourable conditions of climate or food. Black capped chickadees can decrease their body temperature as much as 7°C during the night, thus reducing their energy demands by

60 per cent as the temperature falls in winters; nocturnal moths start shivering to generate heat. Pupfish become more active in the water having 42°C temperature. Some animals show age dependent rhythmicity.

Circadian rhythms may change with the change of place, they may also change with age, e.g. young wood chucks remain active during the evenings only, while their adults, remain active almost throughout the day. In human beings also the sleeping cycle gets shorter with age. Young badgers remain most active during the early hours to the middle of the day, as they grow older, their activity time is shifted from evening to night (Eibl Eibesfeldt, 1950). Similarly differences in the daily activity rhythms of two species occupying the same habitat play important function in reducing their competition, for example fly catching birds and bats never come in direct competition for the same food because birds are diurnal and bats are nocturnal.

Not only behaviour, the biological clocks also synchronise physiological activities with exogenous factors, e.g. the respiration rate of animals is not constant throughout the day, even a small, slowly wriggling creature like earthworm shows a diurnal rhythm in oxygen consumption, peak of oxygen consumption in a twenty four hour period in Lampito mauritii a species of earthworm found in India is associated with its peak activity time (Kulkarni, 1984), the animal was more active in early mornings when there was more oxygen in the environment.

Patra (1984) studied circadian rhythm in oxygen consumption in three air breathing fishes anabas, clarias and channa, and observed a clear rhythm in their metabolism, these fishes were most active during the morning hours, when they moved maximum in search of food and from afternoon onwards they became inactive, or remained submerged in water, in other words these fishes respired more air in mornings and depended on dissolved oxygen afternoon onwards, this rhythmicity is closely interrelated with the diurnal fluctuation of dissolved oxygen. The pure physiological or biochemical reactions within the body of living organisms show rhythmicity.

Vijaylakshmi (1984) studied synchronised shifting of liver glycogen and ATPase activity under varied light regimen in Albino rats. She maintained rats under normal 12 hours light and 12 hours darkness, then in reverse order, i.e. first 12 hours dark and later 12 hours light. She maintained a set of rats under constant light and another set under constant darkness. In the last two situations rhythmic trend in glycogen, glucose and energy linked enzyme ATPase level showed entrainment. Her study indicated the interdependence of the metabolic and related enzyme activity under varied photoperiodism.

By setting the biological clock animals learn to predict environmental events, for example, a nocturnal animal will stop hunting and begin its journey back to den to reach there before the day breaks, by doing so it avoids its own detection by other potential predators, the other obvious example of prediciting ecological changes is seen in the time of onset of breeding, the animals have timed their breeding time in such a way that the young ones are born when it is time for maximum food abundance.

Animals show remarkable time sense; in Panama, pile at ed tinamou is a fascinating bird, it looks like a tiny ostrich, this bird sings every three hours, day and night, in Panama it is called three hour bird. Honeybees have been known to possess marvellous time memory, Auguste Foral, a Swiss naturalist, made interesting observations that every morning when he had his breakfast on the terrace of his house, honeybees came to his table to feed on fruits and jams, their time of arrival was always same every day.

In some cases synchronisation in rhythmicity of certain plants and animals has also been noted. A German scientist, Von Butt el Reepen observed that the bees stayed in his farm around the flowers until about the middle of the day and then left; he further found out that the flower retained same fragrance all through the day, so the bees were not attracted to scent but to nectar which was produced till the afternoon. He gave a new word Zeitgedachtnis (time sense) to express his idea of animals having internal cocks.

Translocation experiments have helped scientists to study the relationship between exogenous and endogenous stimuli. In the eighteenth century French astronomer De Mairan, observed leaf movements in certain plant species with relation to light intensity, he transferred this plant to a dark cellar, to his amazement leaf movements continued for a very long time even in the total darkness. This simple experiment led him to conclude that leaf movements which deemed totally dependent on light intensity were also controlled by an endogenous mechanism which gets them going.

An interesting translocation experiment was carried out in 1960 by an Indian scientist Sirohi together with two US professors Hammer and Hoshizaki. These three scientists went to Antarctica and carried along with them few hamsters, bean plants„ fruit flies and fungi to study rhythmicity in their behaviour. These organisms were placed on a turn table which rotated in such a way that it nullified the rotation of earth which is minimum at poles, the organisms were kept in constant darkness and at constant temperature.

The plants' and animals showed their usual rhythmic behaviour at the pole and there was no influence of earth's rotation either. Hamsters kept their 12 hour sleep wake schedule, bean plants showed 12 hour schedule, the eggs of fruit flies hatched every 24 hours and fungi in petri dish showed light and dark rings of growth every 24 hours. These experiments indicated that once the clocks are set they can go on and on without getting disturbed; but this would not be a permanent feature, depending upon the animal type, species and its behaviour, the clocks will get reset. Circadian rhythms which persists under constant conditions are considered as evidence for an endogenous control mechanism.

The most obviously recognisable daily rhythm in human beings is reflected in "Jet lag" which happens when we cross quickly from one time zone into another sitting in a boeing. Our bodies keep on working according to the time schedule of the zone we have left, and the "clock" will not be "reset" for five to seven days, e.g. there is almost twelve hours difference between New Delhi and New York, when it is 10 am in Delhi, it's 10 pm in New York (of the previous day), after undertaking this journey the day and night are reversed, and the one who travels, would be wide awake in the middle of the night and would feel sleepy during the day. However, this effect is temporary, till the biological clock is reset or entrained. Brown (1970) demonstrated that animals and humans can generate periods other than 24 hours while using a 24 hours clock, for example, a man can get up very late than his usual time on weekends, while still maintaining his usual sleep wake clock for week days, this is called autophasing.

Human body shows clear circadian rhythm in many of its physiological functions too. The body temperature remains high at day time and goes down by night, the blood is less acidic in day time, the adrenals secrete smaller amounts of corticosteroids by night, and the kidneys excrete less urine during night. Many other activities demonstrating rhythmic nature are pulse rate, blood pressure and oxygen uptake in lungs. The intestinal cells when grown in a culture, reveal that they follow the same rhythm of growth, maturation and juice production as the other intestinal cells still within the body forming part of the whole organism.

Understanding of the biological clocks has enabled man to find a variety of its applications, the hunters know when are the animals going to be up and active; the fisherman knows the exact time to harvest marine animals specially the crabs, grunion, palolo and other worms, farmer know what fruit plants will grow in which season, or what are summer and winter flowers and vegetables. There are other more direct useful applications to human beings, e.g. a lady in New York, USA was unable to sleep until about 3 am and then would feel sleepy during office hours, she had to rely on sleeping pills, it was discovered that her biological clock was totally out of harmony with most of us and she was unable to keep the same schedule as we do. She couldn't sleep earlier than 3 am, the doctors asked her to be awake till 6 am, by then she was very sleepy; two days after she was asked to sleep not before 9 am, like this, in a month's time she reached 9 pm, this experiment worked well on her and she started sleeping from 9 pm to 6 am, by shifting the bed time, doctors helped her to reset her biological clock.

The internal clocks, may it be daily, tidal, or lunar deviate little from the periodicity of respective environmental rhythm these internal clocks go either fast or slow. Hence, they have to be reset or re entrained from time to time. The exogenous stimuli are important to start/set and reset the internal clock and once it is done, the clock functions automatically by internal stimuli, e.g. initially a crab living in a burrow on a seashore learns to time the tide, it leaves its deep down burrow to come upto the surface when it is filled with high tide water but gradually it learns to leave the burrow well in advance so that, by the time it reaches the surface, the tide also hits the shore.

This simply means that high tide is an external stimulus or an exogenous entrainer to set the circatidal rhythm in crab and once the clock is set the internal stimuli anticipate the time of the tide and animal reaches the surface in anticipation. Similarly, in many animals length of the day acts as an important exogenous entrainer, it is useful initially to start the clock and later on, to reset the clock; once set and started the internal clocks can function independently of length of the day, e.g. most of the birds will start chirping at 5.30 am when sun rises and would settle down on trees by 7.30 pm when the sunsets in summers, even if there is an overcast and one can hardly see rising sun the birds would start fluttering almost at 5.30 am.

There are questions which remain unanswered regarding biological clocks like, is the property of having internal clocks transmitted through genes, or it is learned at an early stage of development through conspecifics? The single-celled primitive forms responsible for so much of the sea's phosphorescence in the tropics have a 24 hour rhythm to their luminosity. They are bright for 12 hours, and then relatively subdued for the remaining 12 hours of the day. At first, it was presumed that such an action was controlled by sunlight or by temperature but hundreds of generations of these marine algae were kept in constant darkness and they still followed their original schedule of 12 hours bright and 12 hours dim; from such an experiment it would seem that the periodicity is inherited at least in lower animals.

Another experiment conducted by K. Hoffman (1977) supported this fording, he took three groups of unhatched lizard eggs. These were incubated in controlled conditions of light viz 24 hour day (12 hrs. light - 12 hrs. dark); 18 hours day (9 hrs. light 9 hrs. dark) and 36 hour day (18 hrs. light - 18 hrs. dark). After hatching, young lizards were placed under similar conditions; and members of all three groups showed a circadian activity very close to 24 hour day indicating the innate or genetic control over biological clocks even in higher animals. Aschoff (1963) wrote that in most of the animals studied for rhythms, the clocks seemed to be innate or inherited.

Animals which were reared entirely under controlled unchanging conditions showed the presence of an internal clock which remained arythmic and on the slightest exposure to environmental stimuli became rhythmic. For example, drosophila emerge from their pupal cases at dawn and start their 24 hour rhythm. If their eggs are kept under constant lab conditions they will hatch at any time of the day; in an experiment drosophila were kept under constant conditions, for 25 generations, with each generation hatching randomly throughout the day, in one of the generations a light was flashed on eggs, all these eggs hatched 24 hours later, only the single stimulus was enough to synchronise the clocks in this generation of drosophila.

Length of the day or light hours can influence important activities like metamorphosis, which are necessary for survival among animals. The larvae of Sarcophaga and Nasonia develop into pupae and then adults after 14 hours exposure to light. If this period is shortened then the pupa enters into a period of dormancy or diapause for sometime, till it is exposed to desirable day length; on the other hand pupae of many insects hatch during the cooler parts of the day (mornings or evenings), to avoid heat, because their skin is wet and soft, and if it dries up quickly the young insect may even die, therefore, the cooler hours of the day, when the humidity is high and external temperature is low, are favoured so that the skin dries up gradually to become hard.

Biological clock of endoparasites is set according to the daily rhythm of their host animal and the vector. For example, lervae of Wuchereria bancroftii which live in the lungs of human beings, move to peripheral blood vessels below the skin, from where they are sucked by the mosquito, the vector. In Africa, the larvae are transmitted by a nocturnally active mosquito, therefore, the larvae move towards the peripheral blood vessels at night, but in Fiji Islands the carrier mosquito is diurnally active, hence here the Wuchereria larvae travel to the peripheral blood vessels during the day time.

A simple organism like Plasmodium has a highly accurate biological clock. It has precise timings of its recurrent attacks, which are generally multiples of 24 hours. How does this parasite, which lives in total darkness, manage precise timings ? It was found out, that if the host is made to sleep by day and wake up by night, i.e. if the host's normal 24 hour rhythm is inverted or reversed the biological clock of malarial parasite is also inverted, indicating close relation between the two rhythms.

Two main theories about biological clock mechanism have existed (Smith, 1973). One is "The pendulum theory" which suggests that an organism is an independent oscillator, with its own intrinsic timing equalling one day. An ordinary mechanical clock can be an example of pendulum theory. A certain timing has been built into it and that is the speed with which it will tick. The second theory is "Relaxation oscillation theory" which suggests that the organism possesses no such timing but acquires a rhythmic timing from rhythmic and cyclic geophysical events going on all the time. The human heart is a near example. Its speed can be adjusted by its environment, quickened by some hormones, slowed down by others,, accelerated by activity of body and slowed' down by rest. This second theory has been more widely accepted for explaining the functioning of biological clock.

Are biological clocks regulated by biochemical reactions. It is known that temperature greatly affects the rates of chemical reaction, 10°C increase can double the rate of many biochemical reactions, will it also double up the speed of an internal clock? No, it does not happen, even the cold blooded animals whose body temperature fluctuates can run their biological clocks accurately. An animal can have many internal clocks, which are perhaps controlled by a master clock.

The various internal clocks may function independently or in synchronisation. Clunio (mosquito) is found near sea and its habitat is almost always flooded with water except for twice a month, i.e. soon after full moon and also when it is new moon at the time of spring low tide. Only during these two dry times the female lays eggs in the ground. For adult female, the life is very short after it hatches from pupa, soon she must find a male to mate and lay eggs in the ground. The time of hatching, mating and egg laying has to be matched perfectly with the rhythm of the tide. The males, (have wings) emerge a little earlier than the females (wingless) which hatch earlier than the arrival of spring low-tide, lay eggs when the tide reaches its lowest level they die. All this happens within a few hours and only once in every fifteen days. In this case two internal clocks: one clock to determine the right time of hatching and another clock to determine the right time of low tide are functioning in synchronisation.

Where is the biological clock located ? Among lower animals it has been demonstrated by performing careful surgery specially on cockroaches and crickets that these insects can lose their daily rhythm if the connection between Lobi option (part of the brain) and rest of the brain are either chemically blocked or cut. But, in higher animals like rats by destroying certain part of the Nucleus Suprachiasmaticus in diencephalon, the circadian rhythm can be eliminated. In most higher vertebrates it is the pineal body which controls the circadian rhythm.

The functional organisation underlying the circadian rhythmicity is, however, largely unknown, there are definitely many more evidences of presence of biological clock than physiological evidences explaining the rhythmicity. It was observed by Kulkarni and his team in 1984 that in Hirudo birmanica excision of brain disturbed the rhythimicity in oxygen consumption, and when such leeches were given injections of brain homogenates the rhythmicity was resumed. In Crustaceans the circadian rhythms could be altered by removing the eyestalks. It is possible that the substance responsible for rhythmicity may be a proteinaceous neurohormone released through axons (Arechiga, 1977).

Periodic Changes

All the physical factors including light, temperature, with their daily progression makes itself felt in all but it is to be expected that in deep water, soil or cave communities, majority of population will exhibit periodicities that are either directly or indirectly related with the changes occurring during the 24 hours day-night period.

The term diel (day) periodicity refers to all those events which occur at intervals of 24 hours or less. Community periodicity or cycle rhythm results when whole groups of organism exhibit synchronous activity patterns in the day-night cycle. For example some are active only during the night time, i.e. nocturnal, others during the day, i.e., diurnal and still others only during twilight periods, *i.e.*, corpuscular. Singing of bird normally signals the arrival of dawn and butterflies, dragonflies and bees become conspicuous. At dusk, light fades and diurnal animals retires while the nocturnal animals become active.

In both lakes and oceans the vertical 'migration' of zooplankton organisms occurs regularly. It is the striking example of diel periodicity in aquatic habitats. Copepods, cladocerans, larval forms, etc. which make up the vast floating life in the open waters generally move upwards or towards the surface at night and downward during the day-light hours. Whereas in this case light is clearly the controlling factor where, these diel movements are complex and their physiological mechanisms have not yet been fully elucidated. Generally different stages of the same species and each species responds in a different manner so that the crowding of all organisms is avoided into the same region. Although stratification is much more marked at mid-day in comparison to mid-night.

Significance of Seasonal Periodicities

Seasonal periodicities are likewise nearly universal in communities and are generally responsible for almost complete change in the community structure during the annual cycle. In regulating the structure and function of a community, the role of temperature, photoperiod, wet and dry seasons, etc. have been fully established.

Temperature interacts with day length in the temperature zone. Thus Leopold and Jones (1947) found that variation from year to year in the time of flowering of plant and the arrival of migratory birds is greater in early spring when temperatures are critical than in late spring, even though experiments show that the length of day is the controlling factor in many of the cases. Thus in all the insects and herbaceous plants disappear, the summer resident birds return south and winter visitors arrive. The tide rise on the ocean shore and falls about 50 minutes late each day and affects the life on the edge of the sea. The movement of the earth relative to the sun and the moon are underlying these rythmicities.

The alternation of day and night occurs by the earth's rotation on its axis, and the tilt of earth's axis, along with its annual revolution around the sun results in the producing seasons. Orcadian Rhythms (Circa = about; diel - day).

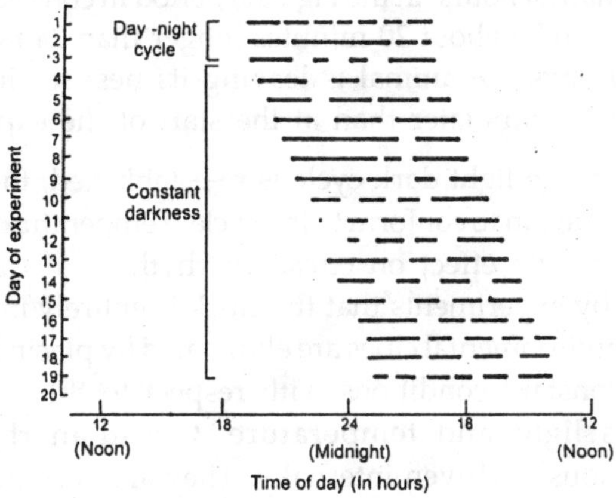

Circadian rhythm in a flying squirrel. Spontaneous activity under a regular light-dark cycle repeat itself each day at the same time, but under constant darkness the rhythm proves to be "free running" in that the activity period drifts, thus indicating that the innate period is approximately, but not exactly 24 hours in length (slightly longer than 24 hours in this example).

The persistent periodicity regulated by the biological clock that couples environmental and physiological rhythm, is called circadian rhythm. It may also be defined as the ability to time and repeat functions at about 24 hours intervals even in the absence of the conspicuous diurnal clue's such as light.

Life is evolved under the influence of daily and seasonal

environmental changes. So it is natural that plants and animals would have some rhythm or pattern to their lives that would synchronise them with fluctuations in the environment.

A nocturnal animal like flying squirrel (Glaucomys) or white-footed mouse (Feromyscus) remains in its nest during the day-light hours and is active at night. The rhythmic activity is exactly 24 hours in length and is entrained or 'locked into' the daily light dark cycle generated by the rotation of the earth on its axis.

If the animal is brought into the laboratory and placed under constant darkness, the daily rhythm continues, but the onset of activity will likely shift slightly each day, indicating a persistent or "free running" period a little shorter or a little longer than 24 hours. In the Fig., the period in complete darkness is shown to be about 20 minutes longer than 24 hours so that after 20 days the animal is leaving its nest for its "nightly" round six hours later than at the start of the experiment.

When the light dark cycle is re-established, the activity is entrained again to conform to the cycle. Temperature fluctuation produces little effect on circadian rhythms. It can never be proved by experiments that the clock is entirely internal since not all environmental clues are eliminated by placing the animal under constant conditions with respect to the major factors such as light and temperature. Circadian rhythms are endogenous or driven internally. They are very little affected by temperature changes, insensitive to a great variety of chemical inhibitors and are innate. They are innate and not learned or imprinted upon the organism by the environment.

Tides and their Applications

Tide is one of the most pronounced periodicities affecting the biota. Tides are produced in the ocean by sun and moon. Since, the sun and its planets exerts an attraction on each other, but the sun, being the largest exerts the most powerful force (Newton's law of universal gravitation).

The earth and the moon also exerts a force of attraction on the same basis. The moon is much closer to earth than the sun and it exerts a force twice as great as that of the sun. Ocean waters exerts two types of bulges by the gravitational pull of the sun and the moon.

Such bulges, tides usually consists of a regular rhythmic rise and fall of water twice a day over a range of several feet, but in shallow waters under special conditions, this sometimes amount to as much as fifty feet. The changes received at the shore line covering and uncovering the tidal zone, directly affect the biota by covering them with water, on by exposing them to air and sunlight.

At the same time, tide occurs on opposite side of the earth. Intervals between successive rises of the tide are about 12 hours, 25½ minutes, or between daily rises about 24 hours, 51 minutes. Because this is the same interval as the average daily passage of the moon across the meridian, it is almost certain that the moon is the principal regulator of tides.

Moon day is 51 minutes longer in comparison to a sun day. In the relationship of moon and sun tides there is constant change, at new and full moon, the attraction of sun and moon on tide reinforce each other, and cause the tides to rise higher at high water and fall lower at low water, without changing the mean sea level. These are known as spring tides. At first and last quarters of the moon, the forces oppose each other and the daily fluctuations are much smaller due to neutralising effect of the force. These are called as neap tides.

Due to the immensely greater distance from the earth, the tide-raising force of the sun is only about 5/11th of that of the moon. When the sun and moon are pulling together, they tend to pull the water of the ocean towards themselves. As a result water is drawn away from the tides of the earth at 90° meridian where the combined pull is most effective and thus produces low tides on both sides of the earth at right angles to the direction of sun and moon.

The tide gradually decline in height, as the sun and moon change relative positions with respect to the earth. When they are pulling at right angles to each other, the sun neutralises 5/11 th, of the moon pull and the tides are correspondingly reduced to neap tide. The theoretical sun tide merely results in lower high tides and higher low tides. Theoretical figures indicate 11 + 5 for height of spring tides and 11 — 5 for height of neap tides, with intermediate fluctuations between the two extremes.

According to the theoretical observations the tides should occur when the moon is on meridian, but because the water cannot follow as fast as the apparent movement of the moon, there is a definite lag of tide behind the moon. In addition, the masses interfere with the ocean tides and cause great irregularity both in the tidal lag and in the height of rise and fall. Whereas the ordinary tide of ocean is usually about 2-5 ft., this figure is often highly multiplied against the shoreline. All of these factors and other minor items like biometric pressure, latitude, on shore and off shore winds, contour of shore, depth of water, natural periods of oscillations must be integrated before predictions of tides can be made at any particular place.

Significance of Lunar Orbit

Movement of moon in its orbit around the earth, causes additional periodicities. Its solid surface not being self luminous is a source of reflected sunlight, some of which reaches the earth as moonlight. Although there is constant reflection from the moon but still its reflection does not reach the earth when it is between the earth and the sun.

Lunar periodicity depends upon the lunar orbit of about twenty-eight days in which the moonlight grows to full light and fades into darkness again, requiring about two weeks for each of these two phases. In the daytime the moonlight is of little significance to organisms but at night the rise and fall of moonlight is important and effects them in several ways.

In comparison to daily and seasonal rhythms, the adaptations to moonlight periodicities are quite less common. Some of the most striking lunar responses are found in marine palolo worms and in the 'grunion fish', in which the behaviour patterns are correlated not only with moon phases, but also with seasons. The pacific palolo in Southern hemisphere rises to the surface to spawn in great numbers when the moon is full during October and November, but the Atlantic palolo in the northern hemisphere spawns at the surface about the last quarter of the moon in July.

The grunion rhythm is much more complicated. The small fish of the species Leurethes tenius come to the southern California coast to spawn from March to July each year. The fish ride the high waves up the beath until they are left stranded on the sand during the nights immediately following high tide of the full or dark moon. Spawning takes place in the sand first below high tide mark and the eggs are burried in the sand. Two weeks later, most of the eggs are ready for hatching and are washed out of the sand by the next high tide.

Biological Clock and its Application

Photoperiod or the day-length period is one of the most dependable cues by which organism time their activities in temperate zones. In contrast to most other seasonal factors, day-length is always the same for a given season and locality. In the annual cycle the amplitude increases with increasing latitude, thus providing latitudinal as well as seasonal cues. Photoperiod has been shown to be the timer or trigger that sets off physiological sequences which bring about growth and flowering of many plants, molting, fat deposition, migration and breeding in birds and mammals, and the onset of diapause in insects.

Photoperiodicity is coupled with what is now widely known as the organism's biological clock to create a timing mechanism of great versatility. In brief, organisms bears a physiological mechanism, for measuring time which is known as the biological clock. Among rhythms, particularly the cercadian rhythm are the basic manifestations of biological clock.

Day-length serves as a sensory receptor like an eye in animals or a special pigment in plant leaves which in turn, activates one or more hormone and enzyme to bring physiological or behavioural response. It is not known just where in this sequence time is measured actually. Although there is wide morphological divergence in higher plants and animals but the linkage with environment photoperiodicity is quite similar.

Similar to plants, animals may also respond to either short or long days. In most of the photoperiod sensitive organisms the timing can be altered by experimental or artificial manipulation of the photoperiod. It is observed that artificial light can bring brooktrout into breeding condition four months early. In migratory birds there is a period of several months after the fall migration when the birds are refractory to photoperiod stimulation. The short days of fall are apparently necessary to 'reset' the biologicai clock, as it were, and prepare the endocrine system for a response to long days.

After late December an artificial increase in day length will bring on the sequence of molting, fat deposition, migratory restlessness and gonad enlargement, normally occurs in the spring. Among birds the physiology of this response in now farely well known, but it is uncertain whether the fall migration is brought about by the direct stimulus of short days or is timed by the biological clock that was set by the long photoperiods of the spring.

In certain insects the photoperiodism is noteworthy because it provides a sort of 'birth control'. 'Brain' is stimulated by long days of late spring and early summer to produce a neurohormone that brings on the production of a diapause or resting egg that will not hatch until next spring no matter how favourable are temperature, food and other conditions.

Thus population growth is halted before, rather than after, the food supply becomes critical.

It has also been observed that the number of underground nitrogen fixing root nodules on legumes is controlled by photoperiod acting through the leaves of the plant. Since nitrogen fixing bacteria in the nodules requires food energy manufactured by the leaves of plant to perform their work. The abundant light, chlorophyll and maximum coordination between the plant and its microbial partner is thus enhanced by the photoperiod regulator.

The process of the biological clock has not been resolved, but two theories have been proposed:

1. *The Endogenous tinier Hypothesis:* according to this hypothesis 'clock' is an internal device that can measure time without environmental clues, and
2. *The External Timing Hypothesis:* according to this hypothesis the external clock is timed by external signals from the environment.

Zeitgebars and their Characteristics

Many organisms exhibit circadian and circannual periodicities that appear to be closely adjusted to patterns of daylength, temperature fluctuation, or other environmental patterns. The mechanism whereby the period of a rhythm occurs, repetitively and coincides approximately with the presence of some external stimulus is called entrainment. All the cues that provide information to animals about periodicity of environmental variables are called Zeitgebars ("time givers"). Zeitgebars are the entraining agents defined as those cyclic environmental cues that can entrain free-running, endogenous pacemakers. Zeitgebars can influence rhythms by effecting both the phase and the frequency.

Considering the example, when the flying squirrel is provided with a daily cycle of light and dark, its activity rhythm generally conforms to the We cycle the onset of activity occurs at about the same time each evening. The squirrel is thus said to be entrained to the daily light-dark cycle. For most endothermic vertebrates thus far tested, the tight-dark cycle is the critical cue for entrainment.

Further, in most terrestrial organisms, the daily light-dark cycle is the Zeitgebar. This is not surprising, since the daily light cycle is usually the most consistent environmental cue in the terrestrial habitat. For animals in the intertidal zone, the web and flow of the tides may be a prime zeitgebars.

Ectotherms such as lizards and insects that are unable fully to control their own body temperatures may use temperature or light cues as zeitgebars. When entrained, they exhibit increased activity under warmer conditions and decreased activity during the cooler portions of the daily cycles.

15

Life Cycle (Flowering Plants)

The life cycle of any sexually reproducing organism commences with the fertilization of haploid sperm and egg. In contrast to the diploid life cycle of animals, the life cycle of plants involves an "alternation of generations" in which there is a multicellular haploid body, these gametophyte, which forms sexual structures, as well as a multicellular diploid body, the sporophyte, which produces sporangia in which meiosis occurs forming haploid spores.

Haploid spores divide by mitosis to form the gametophyte, and the sperm and egg formed by gametophytes fuse to start the saprophyte generation.

The processes of sexual reproduction and seed development in flowering plants, the phylum, Anthophyta, you will note that many processes are similar to those which occur in animal development, such as cell division and differentiation. However, in some significant ways plant growth and development is very different from animal growth:

— In plants, mitosis is restricted to special areas of the embryo and growing plant called meristems. Meristems formed in the early embryo are retained throughout the lifetime of a plant at growing tips.

— Plant cells do not move and migrate. Cells have fixed positions.

Change in plant shape is determined by increases in number of cells and elongation. Cell differentiation proceeds according to the cell's position within the embryo, or within shoot or root meristem.

— Most plants have open growth, so that some parts of the plant are always in an embryonic stage of development, and new parts are added to the plant all of the time (or seasonally, following dormant periods)

— Plants have developmental plasticity. Many cells retain genetic totipotency so that cells can dedifferentiate and become meristematic.

This is particularly useful for plant regeneration when damaged.

— When and where meiosis and gamete formation occur within the life history of plants is different from those events in animals where meiosis directly produces gametes (egg and sperm).

Reproductive organs are found within a flower:

— The ovule is embedded in sporophyte tissue (the ovary)

— The gametophyte is greatly reduced and retained within the sporophyte's sporangia

— Fertilization is "double" requiring two sperm

— One sperm with the egg:

— One sperm with polar nuclei to form a nutritive endosperm

— Seeds are enclosed within a fruit (the ovary and accessory tissues)

The two major groups of flowering plants are the monocots and dicots, which can be distinguished in the following ways:

Monocotyledonae	*Eudicotyledonae*
Flower parts in 3's	Flower parts in 5's (4's)
Leaves with parallel veins	Leaf venation palmate or pinnate
No true cambium	Cambium usually present
One cotyledon	Two cotyledons
Scattered vascular bundles	Vascular bundles in a ring (cylinder)
Sheathing leaf bases	Leaves usually have a petiole
Monocolpate pollen	Tricolpate pollen

Life History of Flowering Plants

Life history of flowering plants uses special terms for sporangia and for gametophyte structures. It's convention. Moreover, sporangia and gametophytes (which are retained in the sporangia) are found in the angiosperm reproductive organ, the flower. Plants lack motility; they are fixed in space. Yet sperm must get to egg just as much in plants as in animals (although some plants are self-fertile).

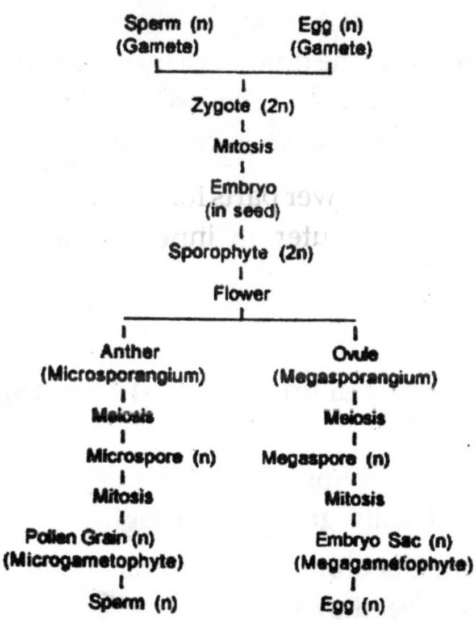

Life History of Flowering Plants

Therefore, plants must be very clever to ensure that one's

sperm gets to a proper egg, and that one's eggs receive proper sperm. To do so, the majority of plants employ a third party as sperm carrier. The functions of a flower are to:

- Attract and reward the "third-party agents", called pollinators or pollinating agents, who transfer sperm to eggs
- Contain sporangia where meiosis occurs, the developing gametophytes (structures which develop from the meiotic product, the spore and which, after maturity, contain egg or sperm), and the gametes produced for sexual reproduction.

Structure of Flower

The flower is a modified shoot system (branch with appendages {modified leaves} attached). The floral parts are attached to the receptacle, the tip of the branch. The flower stalk to which the receptacle is attached is called the peduncle (or, sometimes, pedicel) A group or cluster of flowers is called the inflorescence. Since the flower is a modified shoot it develops from a flower bud located in the axil of leaf. This leaf, which "subtends" the inflorescence, is called a bract.

Flower Parts: The flower parts form whorls or layers within the flower. From the outer to inner whorls the flower is comprised of:

Calyx

- The calyx is comprised of leaf-like structures called sepals.
- The calyx functions to protect the flower bud. The calyx is usually green and tough.
- The calyx is seldom used to attract pollinators; if so the calyx will be showy.

Corolla

- The corolla is comprised of petals.

— Petals are usually "showy" in some way to attract a pollinator. Petals vary in pattern, shape, colour, odour, size, etc.

— Typically, when we think of a flower, we are thinking of its petals.

The calyx and corolla collectively form the perianth of the flower. In some monocots, the petals and sepals resemble each other. They are called tepals, although the petals and sepals can be told apart by their position within the flower. The perianth forms sterile appendages of the flower.

Stamens

— The stamens are the male reproductive structures (modified sporophylls) and comprise the androecium.

A Stamen Consists of:

— An anther, containing the microsporangia (usually four)

— A filament, the stalk that elevates the anther in the flower

Carpels

Carpels (Collectively called a pistil - but that term is not used in this class)

— The carpels are the female reproductive structures or gynoecium (modified megasporphylls folded lengthwise)

A Carpel Consists of:

— A stigma, the receptive surface

— A style, the stalk that elevates the stigma

— An ovary, which contains one or more ovules, the megasporangia

Flower parts are genetically determined by pattern identity genes that are similar to the homeotic genes that determine pattern and placement in animals. Combinations of three genes are important in flower part determination.

Botanists take advantage of this in horticulture to make "showier" flowers, by converting numbers of stamens into petals, for example.

Variations in flower structure are important to the classification of the 250,000 or so flowering plants. Each of these variations has its proliferating vocabulary that students must learn to understand flower structure and flower parts, and is absolutely essential when one wants to learn to identify the different kinds of flowering plants.

The number of floral parts varies in flowers. In monocots, flower parts are typically in 3's or multiples of 3's. In dicots, they are in 5's or 4's or multiples. It is most common to have multiple numbers of stamens. Carpels, and especially the ovaries, typically fuse, and you will read about a 3-parted ovary or 5-parted ovary, which means that the flower has 3 or 5 ovaries that have fused together.

Fusion of flower parts is also common. When petals fuse, we say the flower is sympetalous. Polypetalous flowers do not have fused petals. Stamens and sepals may also fuse. Flower parts can also fuse to other flower parts.

Stamens may fuse to petals. The term, adnate is used when this occurs. Flowers may have all parts or some. Flowers that have all four floral parts are said to be complete. Those lacking one or more are incomplete. Flowers that have both stamens and carpels, the sexual organs, are said to be perfect.

A flower lacking either stamens or carpels is imperfect. Imperfect flowers are either staminate or carpellate. If a single plant has both staminate and carpellate flowers, it is monoecious; corn is an example of a monoecious plant. A plant that produces just staminate or just carpellate flowers is dioecious; examples include holly, Impatiens, and the Katsura tree. Plants that produce perfect flowers are bisexual.

Flower symmetry does not escape special terms. A flower with radial symmetry is actinomorphic. A flower lacking radial symmetry is zygomorphic. Regular and irregular are comparable terms sometimes used. Bilateral symmetry is genetically controlled, sometimes by a single gene, and has developed independently in many floral lines.

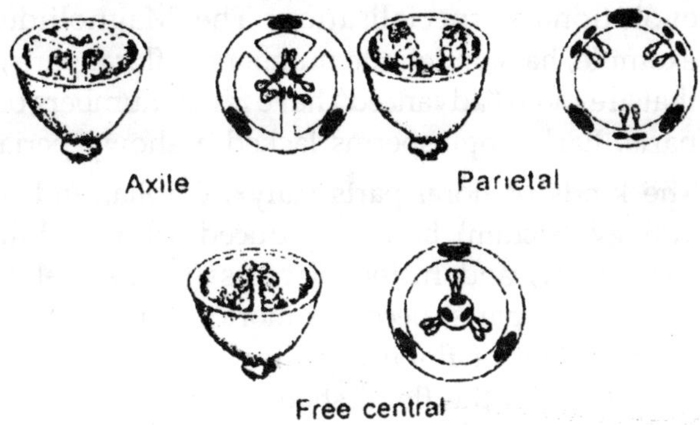

Placentation

The attachment of ovules to the ovary walls, called placentation, is also an important. The attachment position of the ovaries relative to the other floral parts on the receptacle of the flower is also important. Recall that flower parts are generally in whorls on the receptacle from the outer calyx to the innermost carpels.

A flower in which the calyx is outermost and lowest on the receptacle will have its ovaries attached above or higher on the receptacle. These flowers are called hypogynous, and the ovaries are superior. Some flowers have the calyx surrounding the ovaries, forming a hypanthium or floral cup to which petals and stamens are attached. Such flowers are perigynous. Finally the ovaries can be embedded into the receptacle tissue and the other floral parts (stamens, petals and sepals) are attached above the ovaries on the receptacle. Such flowers are epigynous and the ovaries are inferior.

Flower Parts and Evolution: Flower structure and number and fusion of flower parts relates to the evolutionary specialisation of flowering plants. Several trends are evident.

— Pollen with one aperture (monocolpate) is found in the earliest angiosperms. Triaperturate pollen is believed to be more advanced.

— The number of flower parts generally decreases with evolutionary specialisation. The Magnoliides, for example, have indefinite numbers of floral parts; those that are more "advanced" have fewer numbers of floral parts. Early angiosperms lacked a showy perianth.

— The kinds of floral parts (calyx, corolla, androecium and gynoecium) become reduced with evolutionary complexity, and fusion of parts (to themselves and other floral parts) becomes more common. The spiral arrangement of floral parts becomes compacted and shortened on the floral shoot.

— Inferior ovaries are associated with evolutionary specialisation.

— Zygomorphic flower shape is associated with evolutionary specialisation.

Angiosperm Life Cycle

Flowering plants have separate male and female sporangia: the megasporangia and micro-sporangia. Meiosis, the formation of spores, and subsequent gametophyte development occur within the sporangia.

Microsporogenesis and Microgametogenesis occur sequentially within the anther. Microsporogenesis produces the microspore. Microgametogenesis is the process by which the mature male gametophyte develops from the microspore. Anthers contain four microsporangia chambers, often called pollen sacs, or locules.

These chambers contain the sporogenous tissue that undergoes meiosis producing microspores. The sporogenous tissue is surrounded by a layer of nutritive tissue, the tapetum, which delimits the pollen sac. The cells that undergo meiosis are called microspore (or pollen) mother cells. Each microspore mother cell produces four haploid microspores during meiosis. Each microspore develops into a pollen grain that is the male or microgametophyte.

A pollen grain at maturity contains 2 sperm (which develop from a generative cell) and a tube cell. The tube cell produces a pollen tube through which the 2 sperm migrate down the style of a carpel after successful pollination. The pollen grain also develops a thickened bi-layer wall. The outer wall layer, the exine, contains sporopollenin for protection. The inner wall layer is the intine. For sexual reproduction, anthers usually split open, releasing mature pollen that is picked up by pollinating agents and transferred to other flowers.

Megasporogenesis and Megagametogenesis occur within the ovules found in the ovaries of carpels. An ovule is attached to the ovary wall by a structure called the funiculus. The funiculus will also be a seed's attachment to the ovary tissue after fertilization and development of an embryo. It can be seen in some fruits. Each ovule contains layers of sterile tissue, called the integuments and a nucellus layer that surround a single megaspore mother cell.

A small opening between the integuments is the micropyle. The megaspore mother cell undergoes meiosis producing a single haploid megaspore. The other 3 meiotic nuclei degenerate. The megaspore develops by mitosis into an elongate embryo sac (a female or megagametophyte) which contains 7 cells comprising 8 nuclei: the egg, the central cell with its 2 polar nuclei, 2 synergid cells, and 3 antipodal cells. At maturity, the embryo sac is contained within the ovule.

The cells at the micropyle end of the embryo sac form the egg apparatus, and include the egg and the two synergids. The three antipodal cells are located at the opposite end of the embryo sac, the chalazal end. The central cell, containing the two polar nuclei is located in the centre of the embryo sac. The stigma becomes receptive (sticky and shiny) when ovules contain mature eggs.

Pollination and Pollinating Agents

Pollination is the transfer of pollen (containing sperm) from one flower to a second. Since plants cannot move, and sperm in flowering plants is not motile (no flagella), sperm transfer requires a pollinating agent. This dependency poses some problems for the typical plant:

— You must attract an agent

— You must have adequate pollen for transfer

— You must convince the agent to take your pollen to another flower of your species (usually by providing a reward)

— In fact, you want to be so special that the pollinating agent will visit your species exclusively, for at least the time that you are in flower.

— You do not want to produce too much of an attractant or reward because your pollinator must leave you and go visit another flower of your species in order to achieve the goal of pollen transfer. Filling up on you or taking your pollen to a different kind of flower is not useful.

Types of Pollination

— Cross-pollination → Different flower

— Self-pollination → Same flower

Most plants have mechanisms that ensure cross-pollination, which produces more genetic variation. Some plants have back-up self-pollination if no pollinator visits within the needed time frame, and some plants exclusively self-pollinate. Some of the common mechanisms that promote cross-pollination are

— Dioecious plants, and to some lesser extent monoecious plants

— Pollen and eggs may mature at different times, a phenomenon called dichogamy. Protandrous plants have mature stamens first, and protogynous plants have mature carpels first.

- The physical positioning of stamens relative to the carpels can also affect the likelihood of pollen reaching the stigmas of the same flower. The terms "pin" and "thrum" describe the relative position of stamens and carpels in such flowers.
- Genetic self-incompatibility is also common. Pollen grains do not germinate on genetically similar stigmas.

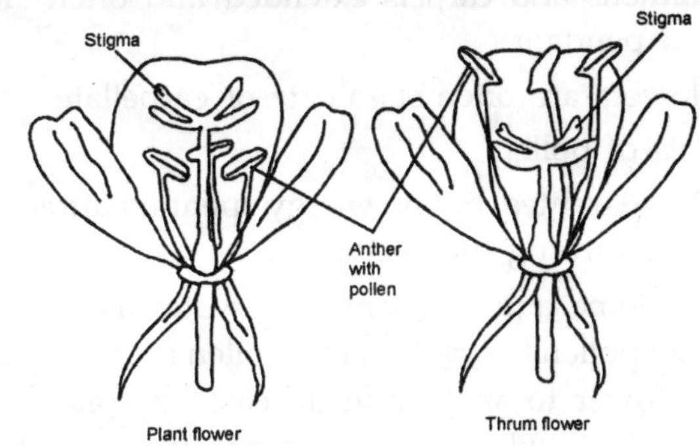

physical positioning of stamens

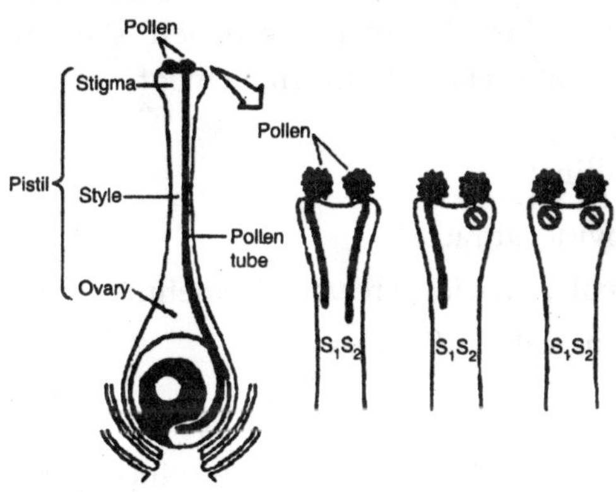

Genetic self-incompatibility

Regardless of the effort to which a plant goes to ensure cross-pollination, the vector, or pollinating agent is still needed. Let's look at some of the pollinating agents and adaptations of flowers for their pollinators. Pollination biology is one of the best examples of co-evolution and "mutualism" in biology.

Pollinating Agents

Wind: Many flowers are pollinated by wind. Many trees in temperate areas are wind pollinated and produce their flowers before their leaves. In general, wind pollinated flowers have

— Reduced or no calyx and corolla to interfere with wind flow

— Stamens and carpels extended and often "feathery" and pendant

— Flowers are often staminate or carpellate

— Lots of pollen

— No guarantee, but no energy spent on attracting and/or rewarding the vector

Water: Some aquatic plants use water as the means of transporting pollen to egg. In some, pollen floats on the surface from one flower to another In at least one plant, the male flower, which is submerged, abscissas from the plant and floats to the surface where it acts like a little boat with sails carrying the pollen. In others the pollen is submerged. Both depend on water currents for movement.

Animal Pollinators

— Provide attractant

— Usually provide a reward (sometimes trick pollinators)

Types of Animal Pollinators

Birds

— Flowers pollinated by birds produce large amounts of nectar.

— Odour is not very important. Birds are visually attracted to the flower.

— Flowers often brightly coloured and large or form large visually conspicuous inflorescences. (red and yellow are most common)

— Flower tubes filled with nectar are common.
— In the western hemisphere, hummingbirds are the most common pollinators
— In other parts of the world, honey creepers and other specialised nectar feeders are found.

Mammals, Especially Bats

— Bat flowers are night blooming, large, often white, and have, for the human nose, an odd odour.
— Bats also eat fruit and pollen. Pollen produced by flowers that is typically consumed by bats is high in protein.
— Bats that eat nectar have elongate tongues to lap the nectar from the flower surface. Nectar is often found in tubes, as well.

Insects: There are three major insect pollinators, Carrion beetles and flies

— Beetle pollinated flowers are either single and large or small and clustered into a tight inflorescence that appears to be a large flower.
— Flower colour is often purple or brownish, resembling carrion
— Many beetle and fly flowers produce carrion odors and attract carrion insects who think the flower is a place to lay eggs. While the insect wanders across the surface of the flower looking for a suitable site, it picks up pollen, which it then distributes to another flower when it tries again.
— Some beetles feed on flowers or specialised food bodies produced by flowers. Such flowers often have inferior ovaries.

Bees, Wasps and Bee Flies: The most famous pollinators are bees. Bees and their flowers have been studied extensively. Bees communicate to other bees locations of good flowers by specialised dances (studied by animal behaviourists). Bees have body hairs and leg (and other) appendages specialised for collecting pollen and nectar, which is taken back to the hive to feed larvae. Most bees must land on the flower to obtain nectar and pollen. It is common for flowers to have designs that guide the bee to the appropriate location and platforms for the bee to land on. These flower patterns are called honey guides and landing platforms. Moreover, some patterns are not visible to humans but are visible in UV light, which many insects can see. Bees do not see red, however, so red flowers are rarely pollinated by bees. Blue and yellow flowers are the most common bee flower colours. Many bees are loyal pollinators and visit just one or a few species of flowers. There are some highly specific examples of bee-flower associations.

Orchids are very specialised for their bee pollinators and often have traps or structures that ensure that the bee reaches where it needs to go. Orchids also concentrate their pollen in modified sticky organs called pollinia that adhere to the bee. One uses the phenomenon of pseudo-copulation as the attractant (and reward). Some bees have become thieves. They have learned that nectar is located in the base of floral tubes (the hummingbird flowers) and chew holes in the base of the tubes to obtain nectar. Those bees that do so, do not pollinate. The native honeysuckle around the BCC campus is a plant subject to robber bees.

Butterflies and Moths

- Butterflies and moths are attracted by both colour and odour, so flowers have both visual and odour attractants.
- Butterflies and moths that serve as pollinators may have a long proboscis, so flowers typically have a long nectar tube. In central America, a flower was found with a tube about 7". A moth with a 7" proboscis was found to be the pollinator.
- Butterflies are diurnal. Many see red, so red flowers are common butterfly flowers.

— Most moths are nocturnal. Moth flowers will be night-blooming and give off heavy, sweet attractant odors.

— Butterflies and moths can hover over flowers and do not need to land on the flower to feed, in contrast to bees and most other insects.

Other Insects: Many other insects are occasional pollinators and nectar feeders. Many flies and mosquitoes are nectar feeders.

Pollen Germination

Successful pollination is just one step in the process of reproduction and development in flowering plants. If pollen of the correct type is deposited on the receptive stigma of a carpel (one which contains at least one ovule with a mature egg), chemicals secreted by the surface stigmatic tissue of the stigma will stimulate the pollen grain to "germinate". A receptive stigma can often be discerned visually, because its surface will be coated with a sugary, sticky secretion.

Pollen germination itself involves uptake of water from surrounding stigma cells by the pollen grain, swelling it until it splits open. As pollen germinates, it grows a pollen tube that grows through the length of the style, through a special tissue called the transmitting tissue that forms a path to the ovary with its ovule(s). The style may have mechanisms to facilitate the growth of the tube. Each ovule must be fertilized by separate sperm, so flowers that have many ovules must also have many germinated pollen grains. For example, each corn silk is a single style.

An average corn on the cob may have 400-500 grains. That is the result of 400-500 pollinations (and fertilizations and embryo developments). Each pollen grain contains two sperm that migrate down the pollen tube. Ultimately, the sperm will migrate to the embryo sac at a region called the micropyle, which is at the base of the synergids and egg. The pollen tube enters one of the two synergids, which has been degenerating during the growth of the pollen tube.

The tube nucleus and both sperm are discharged into the synergid. Chemical signals from the synergid are critical for this function, as are signals and cytoplasmic connections from the egg, central cell of the embryo sac and the sperm cells for successful fertilization.

Fertilization: In contrast to other living organisms, flowering plants require two sperm for successful reproduction.

— 1 Sperm + Egg → Zygote → Embryo Development

— 1 Sperm + 2 Polar nuclei (of the embryo sac) → Endosperm nucleus (Triploid or 3n) → endosperm tissue

The function of the endosperm is to provide a source of nutrients and growth promoting hormones for the developing embryo within the embryo sac. Any endosperm remaining at maturity may be used for germination and seedling establishment. Endosperm may be absorbed by part of the embryo, the cotyledons, which then become "fleshy", to provide nutrients for germination and seedling establishment as well. Legumes are a group of angiosperms that have fleshy cotyledons.

Endosperm is often starchy, but may contain oils and proteins as well. Coconut milk is an example of a liquid endosperm. Following fertilization, the zygote develops into the embryo, the endosperm nucleus into the endosperm tissue, the embryo sac wall and integuments of the ovule into the seed coat and the ovary and accessory tissue into the fruit.

Embryo Development

— The first division of the zygote produces a small proembryo cell and a larger basal cell that develops the stalked suspensor, which elevates the developing proembryo into the endosperm tissue.

— The embryo takes on the shape of an axis with meristems at both ends.
— The meristems are the apical shoot meristem and the apical root meristem, from which structures of the shoot system and root system will ultimately develop.
— In addition, if a dicot, 2 bumps appear near the anterior; these are the 2 cotyledons, characteristic of dicot embryos.
— The cotyledons rapidly elongate, and the embryo is divided into regions, with respect to the cotyledons.
— The region above the attachment of the cotyledons is the epicotyl, which contains the apical shoot meristem
— The region below the attachment of the cotyledons is the hypocotyl that ends with the radicle, containing the apical root meristem.
— Typically the embryonic axis will have to fold, to fit within the embryo sac.
— Endosperm may or may not be absorbed into the cotyledons. It may be consumed completely in the maturation of the embryo, or some may remain for germination.
— Some monocot cotyledons become modified into a coleoptile, surrounding the epicotyl and hypocotyl, and the coleorhiza, surrounding the radicle. This is common in grasses.

Seed and Fruit Formation and Dispersal

While the embryo is developing within the embryo sac, changes are also occurring in surrounding tissues. The ovule integuments and the embryo sac wall form the seed coat, which will protect the embryo until germination. The mature seed is comprised of the embryo, endosperm nutrient tissue and seed coat.

In some seeds, notably legumes, the cotyledons will absorb the endosperm tissue, becoming "fleshy", so that the bulk of the seed is endosperm. In many monocots, the cotyledon will be modified to form a coleoptile, which surround and protects the epicotyl and the coleorhiza that protects the radicle.

In some grains, the cotyledon forms a nutrient-rich scutellum surrounding the embryo for germination. Nutrients also pass from the endosperm through the scutellum to the embryo.

The ovary (and possibly accessory tissue) will form the fruit, which further protects and aids in dispersal (distribution of the embryo to a new location for growth). There are many types of fruits and seed coats, all adapted for dispersal and protection of the embryo. Just as angiosperms require pollinators for transport of sperm, many angiosperms require dispersal agents, and use many of the same attractant and reward mechanisms.

Characteristics of Dispersal Agents

Wind

— Small dispersal unit
— Light in weight
— Appendages to remain aloft
— Tufts of hairs
— Wings

Water

— Air bladders or spongy tissue for buoyancy
— Impervious seed coat

Mechanical

— Pressure sensitive expulsion of seeds
— Humidity sensitive expulsion of seeds

Animal

— Fleshy seed or fruit

Animal consumes and seed passes through digestive tract (seed coat must be acid resistant)

— Attach to Fur or feathers of animal and carried

Attachment devices include:

— Briers

— Spines

— Barbs or hooks

— Bristles

Dispersal is still a chance event. Seed must land in an area favourable to growth, and ideally at a time favourable for growth.

Germination and Seedling Establishment

Many seeds mature at a time unfavourable for growth. Some essential condition for growth may be missing, such as warm temperatures or moisture. In areas where environments change seasonally, and when seeds mature at a "bad" time, most seeds have a built-in delay in germination, called dormancy.

In dormancy, the seed experiences and "waits out" the bad time, and germinates when the surroundings are favourable for growth. Dormant seeds can remain viable for extended periods of time; the metabolic rate is very diminished, the seed coat protects and the moisture content of the seed is excessively low. Dormant seeds monitor their surroundings in a variety of ways, chemical, mechanical and temporal.

— For some, the seed coat must be exposed to abrasion, to physically wear it away.

— Others undergo changes in hormone concentrations, notably abscisic acid and gibberellins.

— Many seeds break dormancy in response to environmental triggers such as temperature changes, fire, drought or light exposure.

Germination is initiated when the seed experiences the "trigger" to break dormancy which is the cessation of the "bad" experience. Most seeds will not germinate until the bad experience has taken place. For example, a seed that must experience cold, will not germinate if harvested and kept warm.

Once dormancy is broken, the seed can successfully germinated if the proper conditions are met.

Conditions needed for Germination (post dormancy)

— Imbibition of much water, to restore the seed moisture content.

— Sufficient oxygen to provide for rapid aerobic cell respiration

— Warm temperatures to facilitate rate of enzyme activity

It also helps if the embryo has a good nutrient supply (often a mixture of carbohydrates, oils and proteins) for cell respiration. Hormones produced in the embryo regulate the conversion of stored nutrients in the seed for germination.

Process of Germination

— Emergence of the radicle, which is positively gravitropic

— Emergence of the hypocotyl (or epicotyl), which is usually arched (from the bent embryo position). The bent hypocotyl protects the shoot axis meristem.

— Upward (or negatively gravitropic) growth of the hypocotyl and shoot axis, which must be rapid. If nutrient reserve is consumed before the emergence above ground, the plant will die.

— The exposure of the shoot to light triggers the development of chlorophyll and also straightens out the shoot.

Rapid establishment of roots for water and mineral absorption and chlorophyll development in the shoot are critical, so that the seedling can provide its own nutrient supply via photosynthesis. If everything is successful, a new plant is established which will continue to exhibit an open growth pattern. It is beneficial to mention two exceptions to normal seed development:

Apomyxis

Seeds are produced without fertilization. Example: Dandelion seeds

Parthenocarpy: Many fruits can be produced without seed development. This process, called parthenocarpy, may or may not require pollination. Certain hormones regulate the maturation of the ovary into the fruit portion of the plant. Either mutation, such as with the navel orange, or the use of hormone sprays on flowers, can result in the maturation of the ovary without seeds. Parthenocarpic plants must be propagated asexually. Navel oranges, for example, are propagated by grafting branches that produce navel oranges onto other orange tree stock.

Plants, Secondary Metabolites and Coevolution: There have been several opportunities in our discussion of flowering plant reproduction and development to mention coevolution in these processes, particularly with pollinating agents and with dispersal agents, two times in the life history of the flowering plant when it beneficial for the plant to provide some of itself to another organism. The same phenomenon, lack of locomotion, that makes it necessary to have "outside" agents for these essential processes can be a negative to the plant during other periods of its life.

The secondary metabolites of many plants are also successful coevolutionary mechanisms to help plants minimise predation. Secondary metabolites include alkaloids, quinines, essential oils, (many of which are terpenoids or terpenes) glycosides, cyanogens, raphides (calcium oxalate crystals) and flavonoids. We call many of these substances phytochemicals, and are investigating them for possible health-promoting benefits.

For the plant, they serve as anti-predator chemicals, either because they are distasteful to the potential predator or cause some other harm. Unfortunately for plants, insects are genetically plastic with much genetic variation for tolerance to nasty chemicals. Some groups of insects specialise on feeding on certain plants. The deterrent chemical may be an attractant to those insects, which suffer no negative impact.

For example, there is a group of insects that feed on cabbage family plants, whose glycosides deter most insects (and many humans, including a former president of the United States). An insect may incorporate the distasteful or toxic substance into its body. The monarch butterfly feeds on milkweeds that contain distasteful alkaloids and glycosides. The monarch adults are distasteful to birds, a survival mechanism for the monarch.

In classic evolutionary success, the viceroy, whose larvae do not feed on milkweeds, and are delicious, morphologically resembles the monarch, and does not get eaten by birds who mistake the viceroy for the distasteful monarch. Wound areas of plants secrete substances that may trigger other parts of the plant to secrete anti-predator substances that will harm the predator if it continues to munch on the plant. In some cases these substances mimic insect growth hormones and alter the growth patterns of the insects.

The efficacy of secondary metabolites has been verified in laboratory situations. Many years ago, slugs were used as test organisms to elucidate the role of nicotine as a predator deterrent in tobacco. Strains of tobacco were grown with varying amounts of nicotine. Mature plants were exposed to slug predation. The advantage of using slugs was the slime trail left behind when the slug "visited".

Without the trail, it would not be possible to know if the plant was rejected, or if the predator just "missed" it. To a slug, they chose low and no nicotine tobacco. They might sample higher nicotine leaves, but rejected them, apparently having more intelligence when it comes to harmful chemicals ingested (or inhaled) than many peopie have.

Classification of Fruits

All fruits may be classified into three major groups on the basis of the number of ovaries and the number of flowers involved in their formation. The following outline includes most of the common types of fruits.

Simple Fruits: Simple fruits develop from a single matured ovary in a single flower. Accessory fruits have some other flower part united with the ovary.

Fleshy Fruits, pericarp fleshy at maturity

a. Berry, consisting of one or more carpels with one or more seeds, the ovary wall fleshy
   - — Pepo (an accessory fruit), a berry with a hard rind, the receptacle partially or completely enclosing the ovary
   - — Hesperidium, a specialised berry with a leathery rind and juice sacs

b. Drupe, a stone fruit, derived from a single carpel and containing (usually) one seed. Exocarp a thin skin

c. Pome (an accessory fruit), derived from several carpels, receptacle and outer portion of pericarp fleshy, inner portion of pericarp papery or cartilaginous, forming a core

d. Hip (an accessory fruit), several separate carpels enclosed within the fleshy or semi-fleshy receptacle

Dry Fruits, pericarp dry at maturity

(1) Follicle, composed of one carpel and splitting along a single suture

(2) Legume, composed of a single carpel and splitting along two sutures

(3) Capsule, composed of several carpels and opening at maturity in one of four ways:

   (a) Along the line of carpel union (septicidal dehiscence)

(b) Along the middle of each carpel (loculicidal dehiscence)

(c) By pores at the top of each carpel (poricidal dehiscence)

(d) Along a circular, horizontal line (circumscissle dehiscence)

(4) Silique, composed of two carpels which separate at maturity, leaving a persistent partition between them

Indehiscent fruits, those which do not split open at maturity

(1) Achene or akene, a one-seeded fruit with the seed attached to the fruit at one point only

(2) Caryopsis or grain, a one-seeded fruit in which the seed is firmly attached to the fruit at all possible points

(3) Samara, a one- or two-seeded fruit with the pericarp bearing a wing like outgrowth. A modified achene

(4) Schizocarp, consisting of two carpels which at maturity separate along the midline into two one-seeded halves, each of which is indehiscent

(5) Loment, having several seeds, breaking into one-seeded segments at maturity

(6) Nut, a hard, one-seeded fruit, generally formed from a compound ovary, with the pericarp hard throughout

Aggregate Fruits: Aggregate fruits consist of a number of matured ovaries formed in a single flower and arranged over the surface of a single recept acle. Individual ovaries are called fruitlets. Many aggregate fruits are also accessory fruits.

Multiple Fruits: Multiple fruits consist of the matured ovaries of several to many flowers more or less united into a mass. Multiple fruits are almost invariably accessory fruits.

Bibliography

Achaya, K.T.: *Indian Dairy Products,* Asia Publishing House, Bangalore, 1974.

Adam, H.: *Food Science,* John Wiley and Sons, New York, 1982.

Addis, Ababa: *Education and Training in Science and Technology in Asia,* UN Economic and Social Council, New York, 1989.

Arbuckle, S.: *Learning of Education,* Allyn and Bacon, Boston, 1965.

Bachman, W.: *Food Chemistry,* Maclaren and Sons Ltd., London, 1955.

Balsdon, J.P.: *Life and Leisure in Ancient Rome,* Bodley Head, London, 1966.

Banerjee, N.: *Indian Females in a Changing Industrial Scenario,* Indo-Dutch Series on Development Alternatives, Sage Publications, New Delhi, 1991.

Cain, M.L.: *Women and Technological Change in Development,* Westview Press, Colorado, 1981.

Carlton, E.: *Foundations of Technical Education,* Prentice Hall, New Jersey, 1964.

Carr, M.: *Appropriate Technology: Two Essays,* Intermediate Technology Development Group, London, 1982.

Crites, J.C.: *Vocational Psychology,* McGraw Hill, New York, 1969.

Donald, E.: *Learning of Technical Education,* Harper Row, New York, 1957.

Dumazedier, J.: *Towards a Society of Leisure,* Free Press, New York, 1967.

Duncan, W.A.: *Engendering School Learning: Science, Attitudes and Achievement among Girls and Boys in Botswana*, University of Stockholm, Stockholm, 1989.

Dyke, S .F.: *The Chemistry of Vitamins*, Interscience Publishers, New York, 1965.

Eley, G.: *Wild Fruits and Nuts*, E.P. Publishing Ltd., Yorkshire, 1976.

Franklin, U.M.: *Will Women Change Technology, or will Technology Change Women?*, The Canadian Research Institute for the Advancement of Women (CRIAW), London, 1985.

Gill, L.S.: *Science, Technology and Gender*, Polity Press in Association with the Open University, Cambridge, London, 1992.

Hanson, J.C.: *Vocational Guidance and Career Development*, MacMillan, New York, 1968.

Helen, Appleton: *Do it Herself: Women and Technical Innovation*, IT Publications, Mumbai, 1995.

Hoyt, K.: *Techniques of Education*, Science Research Associates, New York, 1959.

Hudson, F.: *Development in Food Proteins*, Applied Science Publisher, London, 1982.

James, A.R.: *Education: A System for the Seventies*, Charles E. Marrill .Pub-lishing Co., Columbus, 1970.

Jones, A.J.: *Principles of Education in India*, McGraw Hill, New York, 1963.

Kak, Subhash C.: *Indian Physics: Outline of Early History*, ArXiv, London, 2003.

Khosbed, A.W.: *Education Movement in India*, NCERT, New Delhi, 1963.

Krause, M.V. and L.K. Mahann: *Food Nutrition and Diet Therapy*, W.B. Saunders Co., Philadelphia, 1979.

Krouse, H.: *Counselling Technique with Youth*, Charles and Mervil, Ohio, 1960.

Kumari, P.: *Role of Anganwadi Workers in Integrated Child Development Services Programme*, Discovery Publishing House, New Delhi, 1994.

Lagerstrom, B.: *Indian Women in Industry*, Zimbabwe House (Pvt) Ltd., New Delhi, 1984.

Maclean, M.S.: *Counselling of Technical Education,* McGraw Hill, New York, 1955.

Malcolm, S.: *Science and Technology for People,* Tamil Nader State Council for Science and Technology, Madras, 1985.

Marilyn, Carr: *Technology and Women: Conceptual and Empirical Issues,* Allen and Unwin, London, 2000.

Ndunda, M.M.K.: *Women's Resistance to Science Careers in India,* Queen's University, Kingston, 1990.

O'Kelly, Elizabeth: *Simple Technologies for Rural Women in Bangladesh and India,* Kuber Pub., Dhaka, 1978.

Oblsen, M.M.: *Technical Education,* Holt Rinehart and Winston, New York, 1977.

Pacey, A.: *Culture of Technology,* Cambridge, London, 1983.

Pinto, Diana V.: *Education: Theory, Research and Practice,* Rand McNaIly College Publishing Company, Chicago, 1978.

Potter, N.M.: *Food Science,* The AVI Publishing Co., Connecticut, 1978.

Rahman, A.: *Fatullah Shirazi, a Sixteenth-Century Indian Scientist,* MacMillan, London, 1968.

Rakowski, C.A.: *Women in Non-Traditional Industry: the Case of Steel in Ciudad Guayana,* Venezuela, London, 1985.

Samba, Murthy K.V.: *Science and Technology in Medieval India,* Pratap, Agra, 1982.

Sandler, J.: *The Tech and Tools Book: a Guide to Technologies Women are Using Worldwide,* International Women's Tribune Publications, New Delhi, 1986.

Sharma, P.V.: *History of Medicine in India,* Academic, New Delhi, 1992.

Singh, Harkishan: *Pharmaceutical Education,* Vallabh Prakashan, New Delhi, 1998.

Sponsors: *Proceedings of the Southeast Asian Seminar on Women and Science in Developing Countries,* Kovalevskaia Fund, New York, 1987.

Stadlmare, W.J. and. Cotterill, O.Y.: *Egg Science and Technology,* The AVI Publishing Co., Connecticut, 1973.

Stolte-Heiskanen, V.: *Science: Role of Changing India*, ISSC, New Delhi, 1991.

Subbarayappa, B.V.: *Science and Technological Exchanges between India and Soviet Central Asia*, Educational, New Delhi, 1985.

Theresa, Montini: *Science: Technology and Human Values*, Oxford, New York, 1993.

Tiwari, P.V.: *Kasyapa Samhita or Vrdhajivaka Tantra*, Visvabharati, Varanasi, 1997.

Tolber, E.L.: *Counselling for Career Development*, Houghton, Mifflin, 1974.

Uberoi, Chanchal: *India's Long History of Women in Science*, Physics Today, New Delhi, 1993.

Van, William N.: *Education in the 21st Century*, Houghton Mifflin Co., Boston, 2002.

Veken, M. Van Der: *Women: Technology and Development*, ATOL, Louvain, 1988.

Veldhnis, M.K.: *Food Science and Technology*, The AVI Publishing Co., Connecticut, 1977.

Vijaya Lakshmi, Sharma: *Science in Pre-Modern India*, D.K. Print World, New Delhi, 2001.

Virginia, Hazzard: *Village Technology and Women's Work in Eastern Africa*, UNICEF, Geneva, 1976.

Wardha, B.: *Science and Technology in India*, Centre of Science for Villages, Wardha, 1983.

Whitaker, J.R.: *Principles of Enzymology for Food Science*, Marcel Dekker, New York, 1977.

Winslow, A.: *Science: History*, Duke University Press, London, 1990.

Yukic, T.S.: *Fundamentals of Recreation*, Harper Row, London, 1963.

Zimmerman, J.: *Science in India: Today and Tomorrow*, Praeger Publishers, London, 1983.

Index

S

T

U

V

W

X

Z

□□□